GODLESS FOLLY

Scientific Observations That Refute Materialism

DR. ROBERT GANGE

Godless Folly

Presented by *NewBooks Enterprises*

Cover and Layout design by Laura Shinn
ISBN-13: 978-1484951804
ISBN-10: 1484951808

Scripture quotations are taken from the Holy Bible: The NKJV, NASV, and NIV versions.

Dedication

This book is dedicated to those who have abandoned their folly and are now no longer godless.

Acknowledgements

The author wishes to acknowledge the editorial role and encouragement of Carey Moore, whose literary contributions and suggestions were indispensible in the preparation of this book. His assistance and support is very much appreciated.

The author also wishes to acknowledge the invaluable assistance of Laura Shinn whose formatting contributions proved essential to the preparation of the manuscript files. I am also grateful for her remarkable artistic skills in creating an outstanding cover image.

TABLE OF CONTENTS

Prologue

Two World Views

The question of origins is the most foundational issue facing mankind today.

The reason is that if random events are responsible for human existence, then the following three things are true:

1. We have no purpose beyond this life
2. We are accountable to no authority higher than ourselves
3. We can do anything we want without eternal consequence

Conversely if man is the Creation of a loving, personal God, then:

1. There is purpose to our life
2. We are accountable to our Creator
3. Eternal consequences attend all of our words and deeds

Materialism invites unbridled self-expression.
Theism demands surrendered self-control.

These two world views are separate and distinct from one another. Materialism will be shown to be discordant with accepted observations within the scientific community. It nonetheless remains a popular article of faith among agnostics and atheists. Theism will be shown to be consistent with biblical revelation. This implies that all that we see and are originated from an Intelligence of unfathomable domain and dimension.

The phrase: "all that we see and are" embraces three categories of empirical observations:

i) Expanse of our universe
ii) Advent of biological life
iii) Centrality of human beings

Theism deems all that exists is God and what God holds in being.

Materialism presupposes all that exists is matter and its motion.

The first asserts that a Supreme Intelligence created intelligences i.e. earth now supports about 7 billion people, each is an 'intelligence'. The second believes that non-living specs of dust eventually bounced itself into a living awareness of its own existence to produce such things as pizza parlors, bowling alleys and art museums.

In a very real sense theism holds that life came from life, whereas materialism asserts that life derived from dead dust, that man's ultimate origin is slime — and that when humans physically die, they cease to exist i.e. the individual personality does *not* continue on to another realm.

Thus the question of origins leads us to the question of destiny.

If all that exists is matter and its motion then we have no destiny.

Conversely if we are the creation of a loving, personal God, then our life has purpose and our destiny can be eternity.

The writings that describe this God and His Creation are embodied in 66 Books collectively known as the holy Bible — writings affirmed as such by Jesus Christ, and authenticated when He bodily rose from the grave three days after being publicly put to death. Historical accounts which have survived internal, external and bibliographic testing [1] document at least three eye witnesses accounts and six independent testimonies to His resurrection.[2]

Perhaps the simplest way to illustrate how the Bible relates to the physical world is through the following example: If I design a television set, and then write the repair manual, then both must agree. In the same way, if the same Intelligence authored the Bible and created our world, then both must likewise agree. In other words, the Bible's description of our world must align with what we see and measure. In this sense the Bible is like a jigsaw puzzle cover. It discloses the entire picture when each puzzle piece is put in its proper place.

In reality each of us go through life collecting puzzle pieces in the form of observations and communication. Materialism can be viewed as a scene of the ocean, whereas theism can be likened to a scene of the sky. When we pick up, say, a blue puzzle piece we need to make it fit somewhere in the puzzle. Materialists view it as a piece of the ocean scene, whereas Theists understand it to be part of the sky scene.

To illustrate further, consider the lever arm of a bat, frog and man. Materialism explains these as descending from a common ancestor who had a lever arm (the ocean scene), whereas Theists understand these as an optimum design i.e. as three paintings from one Artist.

The Bible states that our world was created by God for a purpose.

In this sense the Bible tells us 'What' was done, 'Who' did it, and 'Why'.

Differences among believers arise with the questions: 'How' and 'When'.

With respect to 'When', most believers understand the Bible to be silent on the age of things prior to the time of Abraham. There is a small minority which believes things were created only 6000 years ago. But their error stems from erroneously treating Old Testament genealogies as though they were chronologies. [3] The prevalent understanding is that Scripture neither reveals the time of Creation nor the age of the flood.

A second difference arises with regard to 'How' God Created. Some believe God created everything instantaneously with the *appearance of age*. If so, then all of our time measurements are meaningless. It also means God has deceived us into believing that supernovas which we measure as having occurred tens of millions of years ago never existed. Others believe the Bible does *not* reveal how God created, and that for all we know God could have used processes that may have changed through time.

Some question the records themselves and observe that since God is all knowing and wise, why did He leave us with writings whose meaning would allow room for conflict? Others reply that this was done to teach us how to love each other, and that without love everything else is meaningless.

Discussions of this kind notwithstanding, the vastly more important questions are 'Who' did the creating and 'Why', rather than 'How' things were created and 'When'. The reason is that the 'Who and Why' address the very foundation of human existence.

Do we exist as a result of the random motion of matter? Or do we exist as the Creation of a loving, personal God? In attempting to answer this question we will appeal to science — because science is an empirically based system of logic, and logic is that virtue by which truth prevails. To that end it may prove helpful to understand what science is, and is not.

Science is a process yielding provisional knowledge about our world. It's conclusions are never known with certainty. They are only known to a certain degree of confidence. Scientific progress is showing next year that what we believe today to be true is, in point of fact, false; and we spend billions of dollars on it.

Scientific knowledge is not only provisional — it is also synthetic. This means that science only describes events in space and time. Our experience shows that only three kinds of events exist: those that are reproducible, unpredictable, and singular.

Reproducible events lend themselves to scientific inquiry. Unpredictable events lend themselves to statistical inquiry. But singular events lend themselves to *legal* inquiry. Questions of origins e.g., the origin of the world, the origin of life and the origin of man are onetime events. They lend themselves to legal inquiry.

Science has no proper jurisdiction in questions of origins or questions of destiny. Science operates in the present — not the past or the future. But where then does science come in? And what role does it play with regard to origins?

Science gathers evidence to accept or reject the hypothesis of a particular theory. It does this is through a "P H D". The "P" is for Predictions, the "H" for Hypothesis, and the "D" is for Data. However the predictions must be rational, the hypothesis refutable, and the data reproducible.

The predictions are rational when they satisfy the pattern of logical relations in our mind. The hypothesis is refutable when it fails any prediction whose assumptions cannot be satisfactorily altered to make it true. The data is reproducible when other workers at other locations and at other times replicate the data under the same empirical circumstances.

The information in this book highlights established facts from modern science, and applies accepted observations to refute widely held beliefs pertaining to materialism. The preponderance of evidence is seen to dismiss materialism as an unacceptable world view. Materialism views the world as the happenstance of random outcomes, whereas theism presents our world in terms of design by Intelligence. The latter offers us eternity beyond the stars; the former lures us back to the scum of a mud hole.

The question pursued in the following pages is, "What does the empirical data that we find in peer reviewed scientific source literature and published in the professional journals teach? The answer is: Recent scientific evidence has darkened the philosophy of materialism, and instead supports theism as a rational explanation for the origin of our world.

It is noteworthy that our universe would have destroyed itself moments after coming into existence had it not been tuned with a precision of better than fifty decimal places prior to its birth. [4] Apart from the advent of our universe this magnitude of supernatural tuning is unknown in nature.

This and other data discussed in chapter one is concordant with the hypothesis that our world was created by Intelligence — an hypothesis that emerges not because it has been proven to be true but instead because the materialistic alternative is, in point of fact false.

PREFACE

Our Existence

This book — the one you have begun to read, addresses the deepest inquiry that has ever been voiced by human beings. This book probes the question of *existence*. In the past appeal to such an inquiry has been primarily made to philosophy in an attempt to obtain insight into the answer to this question. As applied to *human* existence, philosophy can be understood to be a worldview which cannot be proven to be untrue.

During past centuries two philosophies have emerged regarding the question of human existence. One of these is theism; the other is materialism. To be sure additional philosophical inquiries exist, but these inevitably morph into either theism or materialism. The former asserts, "All that exists is God and what God holds in being." The latter declares, "All that exists is matter and its motion." In its primordial state physical matter is often referred to generically as 'dust.'

This book endeavors to avoid the ideological quicksand of philosophical rabbit trails — and with good reason: Such erudite pursuits eructs an air of higher learning but have cores energized by bias rather than facts. To the extent possible, all conclusions in this book will rest heavily on the latter; bias is averted with a passion.

This isn't to say that I am free of bias — far from it. But the observations made, analysis voiced and conclusions reached embrace factual events, tested methods and logical outcomes. This keeps bias to a minimum. Even so, although everything this book addresses is factual, its content is nonetheless *subject to interpretation.*

This latter restriction is a restraint commonly imposed on scientists. We gather data, ensure it is reproducible, and then use logical predictions to generate an hypothesis which is tested by collecting more data. But the restraint is a blessing in disguise, because it forces us to retain *logical* outcomes. Since logic is that virtue by which truth prevails, retaining logical outcomes is paramount to embracing truth, at least within this context.

These considerations invite the following question: "Can the methods — albeit conclusions of *modern* science be applied to the enigma of human existence? Some believe science can unravel the fabric of physical reality, thereby giving us a portal into materialism. Others contend that the superposed states of quantum physics are the shadow of theism.

These views are but the beginning of a growing realization that modern science may, in some way, lend itself as a tool to cast light on the question raised here, that of human existence. This book cites accepted scientific facts to explore our existence in an effort to bring truth to bear on an enigma that has been pondered for centuries.

This book probes the question of our existence.

CHAPTER 1

Humans and Stars

We begin by observing that relationships exist between human life and the universe. What kind of relationships? One that we will examine is linked through the portal of the carbon atom. However in order to understand the miraculous role that carbon plays in the existence of life on earth, we will need to very briefly examine two kinds of atomic particles called protons and neutrons. For our purposes a proton can be pictured as a red marble; a neutron as a white marble. Six protons and six neutrons combine to create the carbon atom.

All biological life on earth is carbon-based, meaning that if carbon didn't exist then neither would we. So how then did carbon originate? The source of carbon was once quite a puzzle. The reason is that our early universe consisted of only hydrogen and helium— atomic numbers 1 and 2 respectively. Carbon has atomic number 6.

Atomic numbers count the number of protons in an atom— be it hydrogen, helium or carbon. An atom can be thought of as a cluster of protons and neutrons enclosed by a set of spherical wave-like membranes. Each membrane surrounds its corresponding proton but collapses into a very tiny particle (called an electron) when removed from the atom. Neutrons are like protons but without corresponding membranes.

Hydrogen consists of one proton but no neutrons. Helium has two protons and two neutrons. Carbon is composed of six protons and six neutrons. When the universe began there was no carbon— only atoms of hydrogen and helium which, as time passed rose in number due to gravity pulling more of them together. However as they increase in number their mass increases and so too does gravity. This rise in gravity pulls in even greater numbers of atoms which get squeezed tighter and tighter.

Eventually the temperature gets so high (about one billion degrees Kelvin) that collisions among the protons and neutrons kindle a nuclear fire and a 'star' is born. The nuclear reactions initially devour the hydrogen nuclei, and then consume the helium nuclei. So this is how we believe stars come into existence; but where's the carbon? There is none.

However we do know that at temperatures inside stars helium atoms are stripped of their electrons leaving vast numbers of helium nuclei to collide. In doing so, pairs of nuclei momentarily combine to create beryllium (4 protons and 4 neutrons). Although the beryllium is short-lived, (one hundredth of a millionth of a billionth of a second) it nonetheless does occur.

One possible albeit unlikely way to create carbon is for beryllium to first be created, and then for helium to come along, collide with the beryllium, and then stick to it. Since helium and beryllium have 2 and 4 protons, respectively, and since each have 2 and 4 neutrons respectively, when combined they collectively have 6 protons and 6 neutrons i.e. they combine to create carbon.

The problem is that the beryllium must be created and the helium collide and stick all within a hundredth of a millionth of a billionth of a second— the time interval during which beryllium exists. However the beryllium and helium nuclei will only 'stick' to each other if the combined energies of their mass and motion precisely align or 'resonate' with a corresponding resonance within carbon—*a resonance known not to exist.*

In the 1950's Sir Fred Hoyle was researching how stars might create certain heavy elements. In particular his work led him to consider how carbon could be created if only it had the needed resonance— defined by the combined energies of the mass and motions of helium and beryllium (less the small nuclear binding energy).

However instead of seeking an alternate path through which to make carbon, Sir Fred Hoyle reasoned that since we exist, then the needed resonance *must* exist to create carbon. Using the mass and temperatures of beryllium and helium, Hoyle's calculations predicted that carbon must harbor an as yet undetected resonance of 7.6 Mev. Later measurements proved him correct. This is a remarkable result. The existence of biological life led to an accurate prediction of a specific resonance in carbon!

However later research on combining carbon and helium nuclei to create oxygen (8 protons and 8 neutrons) showed it wouldn't happen because the combined energies of carbon and helium were one percent too high. Unlike the nucleosynthesis of carbon in which the needed resonance was found to exist, here the needed resonance for oxygen was absent. However this too is a remarkable result, because had it existed it would have enabled oxygen to be created using carbon as the fuel. This would have diminished carbon to levels so low that it would have disallowed human existence.

Although we may now know how carbon is produced and sustained, we don't know how it is dispersed. In other words how does the carbon free itself from the inside of a star? The answer is the star explodes; not every star—only very special ones that satisfy a host of conditions too involved for discussion here. Suffice it to say that a complex set of reactions occurs over millions of years in heavy stars. Essentially two forces are at play: Nuclear burning of helium within the star's core which pushes outward, and layers of outer material that surround the core and squeeze it inward. Gravity from the outer material acts to squeeze and compress the star's inner core thereby raising its temperature.

The more the core gets squeezed, the higher its temperature and the greater the amount of helium that burns. When there's no more helium to burn the outward push from the core's nuclear furnace stops, and gravity from the star's outer layers squeezes everything so tight that electrons and protons merge together. This transforms virtually all protons in the star's core into neutrons. The conversion is so rapid (under a second) that it invites collapse of the star's outer layer against the neutron core, momentarily squeezing it in a way that sends a shock wave back outward as energy rebounds off the neutron core.

However calculations show that the rebounding shock wave lacks sufficient energy to reach the star's surface and thereby trigger a destabilizing explosion of the star. The additional energy that's needed has been discovered to come from neutrinos (near mass-less particles) created when the star's outer layer collapses against the neutron core. These neutrinos are absorbed by the material in the shock wave as it rebounds off the neutron core. The neutrinos thus give the shock wave the additional energy that it needs to reach the star's surface enabling the supernova explosion to occur.

The absorption of the neutrinos by the material within the shock wave is governed by the 'weak interaction' force. If this force is too strong the neutrinos will never leave the neutron core and reach the rebounding material; if it is too weak the neutrinos will reach the rebounding material but not be absorbed by it. In a manner similar to the resonances discussed earlier, "It has to be just right!"

The next chapter opens by summarizing all of this in a simplified manner, one that illustrates how three independent albeit remarkable interactions of carbon within stars across the universe enable human life to exist.

CHAPTER 2

The Anthropic Cosmological Principle

Let's summarize the remarkable role played by carbon in enabling human life to exist.

* Carbon is produced within certain stars because the total energy of beryllium and helium (their mass, motion and nuclear force) precisely aligns with a resonant energy in carbon.

* Carbon is sustained in stars because no resonant energy exists in oxygen that matches the total energy of carbon and helium (mass, motion and nuclear force).

* Carbon is dispersed throughout the universe because precisely tuned forces of weak interactions orchestrate complex nuclear reactions which lead to a supernova explosion.

These three independent relationships are truly remarkable. They occurred billions of years ago, and billions of light-years removed from earth and yet they enable us to exist. The word element 'anthrop' stems from the Greek word *anthropos* and denotes 'man' or 'human being.' Conversely cosmology is a branch of philosophy pertaining to the origin and structure of the universe, its elements and laws.

Ordinarily we regard the two as disconnected. Man is on earth — a tiny planet located billions of light-years away from the nuclear reactions we have been discussing. Yet when we realize that our existence depends on resonances and forces that occurred billions of years ago at a distance of billions of light-years from earth, we pause to ponder, "Is our universe a gigantic womb designed to spawn and harbor human beings?

There are some who believe that the answer to that question may be 'Yes.' Those who do are not priests and pastors and poets. They are scientists. These researchers are being motivated to believe that something else may be at play by the self-evident relationships between man and his cosmos.

It is noteworthy that in examining carbon we are just looking at one small piece of a much larger pie, Let's consider one aspect of the crust of this pie— the *space* where we live and breathe and walk.

We exist on earth, a planet that orbits our sun. What's important here is that earth's orbit be stable. Were debris from outer space to impact the earth (and it often does), it could ever so slightly move our planet either toward or away from our sun depending on its angle of entry, thereby destabilizing earth's orbit. Fortunately, however, earth's orbit remains stable due in large measure to gravitational forces whose strength changes inversely with the square of the distance.

It is this feature of the gravitational force field that enables us to live on a planet with a stable orbit — a feature that only exists because we live in a three-dimensional world. Were we to live in a space with dimensions other than three we would be unable to exist. The reason is that orbits in a central force field are stable only when the force varies inversely as the square of the distance. So again we are led to ask the question: "Has our universe been designed for us to exist? The assertion that our universe was created to enable our existence is referred to as the 'Anthropic Cosmological Principle.'

John Gribbin and Martin Rees wrote a well-reasoned 303 page book in support of the Anthropic Cosmological Principle entitled *Cosmic Coincidences.* [1] The book cites a number of 'cosmic coincidences' and the role each plays in underpinning the relationships of mankind with his universe.

John Barrow and Frank Tipler likewise wrote a well-reasoned albeit highly technical 706 page book entitled *the Anthropic Cosmological Principle* [2] This book is an in-depth comprehensive examination of the question of human existence in relation to our universe, and shows that there is a connection between the stars and galaxies and the existence of life on planet earth. It also reveals that such life requires a universe with an expanse of billions of light-years.

Implicit within the Anthropic Cosmological Principle is the realization that were just one of the fundamental dimensionless physical constants to be altered by just a few percent life as we know it could not come into being. [3] In the words of John Wheeler (Center for Theoretical Physics, University of Texas at Austin), "According to this principle, a life-giving factor lies at the center of the whole machinery and design of the world."

If this be so, then it means that there is a proper distinction between human beings and their physical world — an epistemological difference between the things we measure, and the observer who does the measuring — a fundamental divergence between the human mind, and the brain tissue through which that mind expresses itself — an ultimate dissimilarity between physical matter, and the intelligence that organizes it to useful ends.

These distinctions are the antithesis of materialism which asserts all that exists is matter and its motion. Materialism views the human mind as a characteristic of physical matter and regards intelligence as an attribute of the motion of its particles. Moreover lurking at the core of this dichotomy is the question of *existence*. We have two options:

> 'All that exists is God and what He holds in being'
> — or —
> 'All that exists is matter and its motion.'

Recent debates on origins, now framed (loosely) in terms of "evolution" versus "intelligent design," have given rise to adherents on both sides of the issue — each side seeking to persuade the other either by the logic of its arguments or, at the very least, through a defense of the position that it holds.

The purpose of our discourse is to suggest that the resolution of the issue lies in a realm deeper than the level at which the debate has been framed, that the epistemological root of conflict has more to do with values than with logic, and that the tension between evolution and intelligent design as competing views to explain origins stems from a deeper conflict spawned by the question of existence.

The question of existence is the question, "Why does anything exist?" and, more narrowly, "Why do we exist"? The question breeds two auxiliary questions: "What *is* existence?" and: "How is it expressed"?

CHAPTER 3

Discourse on Existence

What is existence? Can things exist in an absolute sense? If there is no living agent to bear witness to such things do they really exist? Ultimately, existence is tied to 'one's awareness of being.' Descartes put it well: 'I think, therefore I am'. [1] Stated differently, in the absence of a living agent to bear witness to the data of his or her experience, nothing can be said to exist — regardless of how strongly one may believe that it does.

How is existence affirmed? The answer is — through the body. We communicate and thus know one another because of the energy that is exchanged between our physical bodies. The meaning we impute to the "reality" we recognize as living agents is made possible because of the energy exchanged between — and then processed within our bodies.

Our eyes are cameras — our ears are microphones. These are a subset of a collection of sensors that assemble and transmit energy in the form of waves that reach interpretive centers in the brain. Upon arrival these are assigned "meaning" by the 'living agent'.

The question: Why does anything exist? —has its ultimate repose in the meaning we assign the information processed by our brain. However there is an essential distinction between our brain processing information, and the existence of we as living agents assigning that information *meaning*.

If Intelligence authored the information we receive through the energy exchanged via sensors within our body, then "God" exists, acts in and of Himself and may even bear Witness to our mind as He chooses. However if serendipitous configurations from random permutations give rise to such information, then physical matter and its motion exist in and of itself, without intellectual support existing external to that matter.

The next chapter begins to probe the validity of materialism versus theism by confronting us with three enigmas: *Concealment, Consciousness and Color*. These three attributes are each discussed against both world views: Theism — which holds that all that exists is God, and what God holds in being, and: Materialism — which holds that all that exists is physical matter and its motion.

CHAPTER 4

Materialism's Three Enigmas: Concealment, Consciousness and Color

Two world views confront us:

1. Theism — which holds that all that exists is God, and what God holds in being.
2. Materialism — which holds that all that exists is physical matter and its motion.

Which is correct? They can't *both* be true.

Concealment

Let's assume materialism is correct. If so, then you — the reader and I — the writer each consist of *only* physical matter and its motion. But since physical matter interacts with all other matter throughout the entire universe, and since such interaction appears to be absent in the private experience of human beings everywhere on earth, it suggests that human beings are more than just physical matter. This is explored further in **Appendix 1** (page 83).

It by no means *proves* that materialism is false. But it does provide a hint to this effect, because it raises the question: How can matter interact over the space of billions of light-years [1] and yet show no sign of doing so between two human brains just inches apart?

Consciousness

Consider materialism in the light of several private human experiences. Materialism asserts that nonliving specs of dust eventually configured itself into people who experience things, such as the colors in a rainbow, the affection of a loving embrace, the horror of unbearable pain, the ecstasy of indescribable pleasure and the role of conscience in making moral decisions. It affirms that dead dust eventually enabled itself to build pizza parlors, attend art museums, visit concert halls and play bowling ball.

Each of these in and of themselves raises serious questions. For example, is it sensible to believe that nonliving specs became living people? Is it reasonable to suppose that dead dust created live music? Is it sane to presume that inert grime came to eat and enjoy pizza — or that lifeless grunge eventually came to throw large balls down bowling alleys?

Taken at face value, these things are absurd. How then did such beliefs arise? Certainly *not* from anyone seeing dead dust change into people! No. They arose because large numbers of people refused to believe that theism is true. And if theism is false, then materialism *must* be true, because we have no other alternative [2].

All we've got is theism and materialism. If theism is false, then materialism must be true — and if it's true, then we're led to the absurd conclusion that dead muck morphed itself into gifted artists, great composers, brilliant scientists and even theologians!

But if materialism leads us into absurdity, why is it so welcomed throughout the world?

Why is theism so quickly dismissed? The reason is that Theism avows the existence of an Intelligence that is vastly greater than human intelligence. It teaches that there is an Authority higher than man, and to Whom humans are beholden. But this is disturbing to most humans. Our nature is the propensity to satisfy desire with the least possible pain. We desire maximum pleasure and minimum pain. Humans seek pleasure, and avoid pain.

Why should I live my life the way Someone else wants me to live it — especially if doing so diminishes my pleasure and brings me pain? I want to be king of the mountain. I want to be the highest authority! It's called 'secular humanism.' *Secular* means no God, and *humanism* conveys, 'Man is unassailable. However, theism interrupts all of that.

Theism asserts that God is on top — not man. And if God is on top, then it means He can (and does) command us to live the way He wants us to live, and not the way *we* want to live. And the way He wants us to live defers pleasure to a future time, and invites the prospect of pain into our present existence.

In a word, the way He wants us to live is unacceptable, because it requires us to behave in ways contrary to our nature. And so we are left with two alternatives: Do what He says — or deny that He exists. Since we will not have that seem true which contradicts our passions and affections, man denies God's existence by choosing *materialism.*

At first glance mankind's choice may appear acceptable. But it begs the question: 'Is there an objective reality that speaks to the validity of theism, and the sham of materialism?' That question takes on renewed meaning when we consider 'color.'

Color

Consider the colors of the rainbow cited earlier. When we close our eyes and "see" red or green or blue, we are experiencing an aspect of reality beyond anything that can be expressed, explained or illustrated to a person born blind. No physical description is possible because no physical description exists. It is *impossible* to convey the sensation we experience of red or green or blue to a person blind from birth.

But if materialism were true, then the optical sensation we experience as 'green' would be a physical property of matter that should, could and would be describable. One reason we cannot describe the sensation humans experience as "green" could be that it resides in a realm outside physical reality. Be it color or love, music or art, pleasure or pain — these and other experiences appear to dwell in a realm that is inaccessible to nonliving objects.

Again, this by no means *proves* that materialism is false. But it does provide a hint to this effect, because it raises the question: "If the sensation we *experience* and call "green" resides only in matter and its motion, then why does it not lend itself to being described, as do all other properties of physical matter?"

This 'green' experience is *so* private, that it's not possible to verify that the color sensation I experience is the same you experience when we both look at, say the leaves of a rose bush. All we *know* is that we agree to call whatever color we experience "green."

For all we know, the color *you experience* in looking at the green leaves of a rose bush is the *identical color that I experience* when I look at a red rose pedal — and vice-versa. Neither one of us necessarily experience the color sensation experienced by the other.

All we know is that we each agree to label as "green," whatever color we 'see' *when looking together* at the leaves of a rose bush. We also each agree to label as 'red' whatever color *we both 'see'* when looking at a rose pedal. However, the actual color sensations we experience are *so* private — as to deny us access into the experience of the other. Note: The reader should exercise care not to confuse the electromagnetic waves with the *experience* of the 'color' sensation that we *observe* and to which we each bear witness.

CHAPTER 5

Materialism - and Natural Laws

This chapter introduces the conflict between materialism, and the scientific laws that govern physical matter. However we won't need mathematical equations to highlight the issues. Instead we will illustrate the clash between materialism and modern science through some simple examples.

Materialism asserts, 'All that exists is matter and its motion.' All well and good. But let's now apply scientific laws to describe the location and motion of this matter. We'll start by throwing a ball down a bowling alley. How many pins will our bowling ball hit at the end of the alley? Well that depends on how fast and in what direction the ball is rolling.

Imagine seven people who throw seven balls — one down each of seven alleys. At the end of each alley is a set of pins. Each of seven balls rolls toward the set of pins at the end of its alley. But as each ball collides with its pins, it does so at a speed and direction that differs from the other six balls. Each ball has a different speed and direction! Why?

Here's the reason: The rate that each ball rolls down its alley is determined by the force used to throw it; its path results from the direction given the ball when it left its bowler's hand. Since the force and direction given each ball differs from bowler to bowler, so too does the speed and path of each ball when it collides with its pins.

This trivial example illustrates a foundational albeit universal truth: Neither speed nor path is an intrinsic property of a ball in motion. These instead arise from initial conditions imposed at the onset of the motion. Throw a basket ball toward a hoop. Its speed and trajectory depend on the force used to throw it — and its direction as it leaves the player's hands.

Throughout the universe physical matter obeys scientific laws — but these laws exist in differential form. This means that when we go to solve them, we (humans) need to provide the initial conditions.

If you are mathematically minded, these are the integration constants that must be provided before these equations can describe the actual speed and path that the moving object will take. Scientific laws tell us how physical matter moves; but initial conditions are necessary to describe its observed path and speed.

Human beings are living agents who bear witness to the data of their experience. And as living agents we are free to choose one set of initial conditions, or another. For example, I am free to throw a bowling ball down an alley with all my might, or to gently place it on the floor and push it with my foot. That choice is mine to make.

And the initial conditions I choose (initial speed and path of the ball) do not belong to the ball. No! They are mine to decide — and mine to implement. Moreover once specified, the ball must obey the scientific laws that govern its motion and direction. The same is true of the basket ball.

The universal observation that physical matter obeys natural laws initialized by conditions freely chosen by an observer presents a formidable conceptual obstacle to materialism. If I consist only of physical matter, how is it possible for me to impose on physical matter — initial conditions that are not a property of the physical matter of which I am solely composed? How can I as a living agent consist only of matter and its motion, and yet create initial conditions that are not a property of matter and its motion?

But these questions invite a more perplexing question: 'How can physical laws operate within regions of deep space where there are no humans to provide the initializing conditions? How can the physical matter in these regions obey natural laws that are void of the conditions necessary to initialize their outcome?

In and of themselves natural laws tell us how matter will move. But these laws require initial conditions before they can describe where the physical matter will go. To illustrate, our natural laws reveal that a bullet fired off a high mountain can follow any one of many thousands of 'parabolic' paths. These parabolas describe how the bullet can move.

But where the bullet goes depends on the one parabolic path that the bullet actually takes. What causes the bullet to choose this unique path? The answer is, 'The speed and direction given the bullet as it leaves the barrel.' This 'speed' and 'direction' are the initial conditions needed by natural laws to fully describe physical matter and its motion. In this example a human chooses the amount of gunpowder to use, and where to aim the rifle.

But what of deep space where no humans exist to provide these initializing conditions? We know that our natural laws operate in such regions. But how do they know where the matter will go without the initial conditions they need to describe the physical matter?

Consider, for example the natural laws that describe how stars form. Recent discoveries regarding stars with very small amounts of metal {Hyper-Metal-Poor (HMP) stars} indicate they were formed from gases chemically enriched by earlier stars (called 'supernovae') that implode and then explode in special circumstances.[1] Here chemically enriched gases can be viewed as the initializing conditions to form HMP stars.

But how did these gases arise? Answer: From imploding and exploding supernovae. But how did the supernovae first arise? Answer: From the gravitational inward pull of the star's core becoming stronger than the outward push from the star's nuclear furnace. But how did gravity and nuclear reactions arise? What initial conditions attended these?

Well by now you get the point: Prior initial conditions attend every answer! Every "How did {whatever} arise?" question is answered with events constituting the initial conditions for the {whatever}. If we ask, "How did gravity arise?" the answer is, "From the advent of mass." If we ask "How did mass arise, the answer is "From curvatures in space-time."[2]

The amount and position of mass define gravity's strength and location. These mass parameters are gravity's initializing conditions, whereas the strength and breadth of space-time curvature provide initial conditions needed to describe gravity. As we regress further back in time with our "How did {whatever} arise" questions, we are driven to an initial condition for which only theism has an answer.

This 'primal' condition is the inception of our universe, commonly called "Big Bang," and when we ask, "How did it arise?" theism's answer is, "God." Materialism has no answer because matter, its motion and even time itself come into existence at this point.

But human experience teaches that 'primal' initial conditions are created by intelligence. We choose a path and speed to throw a ball, or gunpowder in a bullet and where to aim it. In asking, "How did the ball go through the hoop?" we need go no further than its speed and direction when thrown. Since this is true of all events on earth with humans, should it not also be true of cosmic events with Big Bang originating by Intelligence we call God? Does the miraculous precision of our world's advent leave us another rational option? [3]

Chapter 6

Initial Conditions Are a Crucial Requirement

The revelation that natural laws require initial conditions before they can describe where objects go carries an implication that severely compromises materialism. All events have their ultimate origin in some 'primal condition.' This is true of even random events, in that something or someone is ultimately responsible for setting them in motion.

For example, any object immersed in a liquid (or gas or solid) undergoes random movements called 'Brownian' motion. This results from collisions between the object and the gas molecules. The random motion occurs because these molecules are in thermal motion due to (in this case) the 'primal condition' of heat energy.[1]

In the case of human events, such as a car moving down some street, the primal condition is the person turning a key to start the engine. In the case of natural events, such as an earthquake or volcanic eruption, the primal condition ultimately traces back to Big Bang.

But consider that all human events ultimately involve an act of free will. A window washer falling to the ground because his scaffold broke still had to first freely choose to go onto the scaffold. If he was forced to go onto the scaffold, then someone freely chose to force him onto the scaffold. Pickpockets freely choose to steal wallets from others.

The point is: "All human events ultimately trace to a free choice of will. This free choice of will is the primal condition that initializes events. The self-evident fact that we are free to choose is troublesome to materialism. The reason is that if human beings only consist of matter and its motion, then the laws that govern physical matter ought to also govern human beings. But if this be so, how can humans possess the liberty to choose freely? How can we be free to override laws that govern us?

The materialist attempts to escape this noose by dismissing man's freedom to choose as a deception. However, what he has actually done is denied the reality of human experience! The committed reductionist alleges, "Humans only think they have free will; but they delude themselves. They allege, "There really is no such thing as free will. It's an illusion." Here the materialist metaphorically portrays man as a stone that is thrown high into the air which, at the highest point of its climb says, "Shall I go up or shall I go down? I choose to go down.

But a materialist observer says,

"The stone must go down! It has no choice. It may believe it went down because it chose to do so. But I know better! I'm sure its freedom to choose is an illusion."

The materialist continues,

"The stone is subject to a gravitational force field. Throw as many stones as you like up into the air. They must all come down. No matter what choice each stone may think it has — gravity will make it fall to the ground."

Those committed to materialism echo the reductionist mantra by alleging:

"Man may think he has freedom to choose. But like the stone it's an illusion. Humans erroneously believe they create initial conditions that are not a property of physical matter. This too is an illusion."

The foundational point to note here, however is that this feeble attempt to perpetuate materialism rests upon denying the self-evident observation that we are free to choose.

Is it possible that everything is an illusion? Yes — but that's because in synthetic thought anything is possible. It's possible that you reading this book is a dream. It's possible I'm an alien from Andromeda sent here to deceive you! The question we must ask, however is not 'Is it possible?" but, "Is it true?"

The foundational point is that whereas matter moves along paths determined by natural laws, how fast it moves and where it goes is dictated by initial conditions freely created by humans. That's an empirical fact. Therefore we cannot be governed by the laws that govern physical matter — as would be true were we to consist of only matter and its motion. Were we governed by the laws that govern physical matter, it would not be possible for us to create initial conditions that were independent of such natural laws.

CHAPTER 7

Transitioning from Classical to Quantum

This is a short chapter but an important one. The word 'chafe' means 'to warm by rubbing.' And that's exactly what happens each time we observe or measure something. The material 'stuff' we look at gets slightly warmer due to our looking at it.

It's amazing, however, that we didn't realize this occurred until early in the twentieth Century. Up until then we believed that the stuff out there had an external existence apart from us, and that when we measured something, the thing being measured had no way of knowing that we were looking at it. But in point of fact this is not true!

Let's quickly assess what occurs and see how it leads to the kiss of death for materialism. Any time we see anything, it's because light energy enters our eye and hits our retina creating electrical signals that travel along neurons into our brain.

But where did this light energy come from? It reflects off the object we are looking at. For example, on a bright sunny day light from the sun comes to earth, bounces off the apple and enters my eye. That's how we see the apple.

We can see the apple in a dark room by shining light from a flashlight onto the apple. Just like sunlight, the light energy from the flashlight hits the apple and bounces into our eyes.

But regardless of whether light energy comes from the sun or a flashlight or whatever, this light energy comes to us in the form of packets of energy called 'photons.' These photons can be thought of as ultra tiny, infinitesimally small bullets of light energy.[1]

Now these photons are so small that when they hit and bounce off an apple, the apple is virtually unaffected. It's like hitting an aircraft carrier with a feather. But when photons hit things that are themselves also very tiny such as an atomic particle, a small part of the photon's energy causes the tiny things they hit to move, thereby raising their temperature ever so slightly.

Here the object observed acquires kinetic energy (energy of motion) that it gets from the energy in the light. This small loss of light energy causes the light to change color toward the red (the photon's wavelength gets slightly longer).

One of the great 20th Century discoveries is that it is impossible to measure anything without an exchange of energy occurring between the observer, and the thing observed.

Before we knew of this energy exchange, our scientific laws described 'things out there' without any reference to the observer. These are now called "classical laws."

CHAPTER 8

Materialism's "Quantum Kiss of Death"

When we learned that observing something very small affected its behavior by transferring energy to it, we no longer were able to describe its behavior without also including the energy exchange between it — and us as observers. The observer was now part of the system being described.

These new laws are called "quantum laws." As you might expect, these 'quantum laws' morph into 'classical laws' when the object observed becomes very large. In this case the energy transferred is so small in relation to its mass that the impact is, for all practical purposes negligible (but not so for very small objects).

Things that obey quantum laws are called 'quantum objects,' and things that do not obey quantum laws are called 'classical objects.' In the world of quantum physics, the things we observe are called 'quantum objects' because they obey quantum laws. Conversely the observer is often referred to as a 'classical object' in that s/he is exempt from these laws. The reason is all quantum laws are statistical relations among human observations. [1]

This 'exemption' of observers from quantum laws gave rise to the following mind-body question: "Does an observer exist apart from the objects s/he observes? We can rephrase this question as follows: "Do human observers consist of only matter and its motion — or do we consist of something else, apart from matter and its motion?

But here it's important for the reader not to confuse the human body (quantum object) with the agent who bears witness to the data of his or her experience (classical object). Now it may be that the answer to the mind-body question could lead us to conclude that we (agent) are our body, in which case we would only consist of matter and its motion.

So here's the question we need to answer: Are we as living agents nothing but our body — so that we go out of existence when we die? Or are we as living agents something apart from our body — something other than matter and its motion?

Can this question even be answered? Not only can it be answered, but in point of fact it was answered years ago by Eugene Wigner [2], a world renown quantum theorist and past colleague of Albert Einstein, and a Noble laureate [3] in whose honor international conferences in quantum physics have been convened (among many other honors). [4]

Wigner offered an ingenious insight to resolve the question. However for me to explain what he proposed and how he managed to answer the question requires that we simplify things even further. We do this so that a reader not trained in quantum physics can understand what proved to be the resolution of the deepest question mankind has ever pondered viz. do humans consist of only matter and its motion, or is there more than this?

We're going to answer this question by explaining the experiment Wigner proposed in terms of things familiar to the reader viz. Coke and Pepsi. Now we all know what these are. Each basically consists of the same kind of stuff i.e. soda. But we're also familiar with oil and water. These are not the same kind of stuff. In fact they're so different that when put into the same glass one floats on top of the other.

Now let's recall that it is impossible to measure anything without an exchange of energy occurring between the observer, and the thing observed. This means that the total system being described must include the observer along with the object being measured.

Both are key components of the system, because the energy that the observer needs in order to observe the object impacts the object and thus alters its location and speed in space. Thus in order to describe the complete system, one must include both the observer and the object s/he observes.

In quantum physics the observer (along with the energy needed to measure) is characterized by a mathematical entity called a 'wave function.' Likewise the object being measured (along with the energy impacting it) is also characterized by a wave function. [5]

Since the observer and this object differ from one another, and since each is located in different parts of space, the parameters that define their wave functions likewise differ. We can think of these wave functions as mathematical pictures of a deeper reality where the "stuff" that composes these objects resides.

Please understand that we are vastly simplifying things here for purposes of illustration. The actual quantum expressions needed to describe this system along with the calculations required to solve the relevant equations are extremely difficult to express.

Wigner's great genius is that he not only conceptualized the relevant quantum mechanical framework and developed the needed mathematical formalism, but that he actually solved the mind – body problem that he conceptualized!

What is the 'mind – body' problem that Wigner conceptualized? We have said that the observer doing the measuring, and the object that is being measured both belong to the same (joint) system. Quantum physics describes each of these in terms of wave functions.

Thus we start with two wave functions — one for the observer, and the other for the object being measured. And when the observer takes a measurement, an exchange of energy occurs between the observer and the object that forces the two wave functions to interact.

But how they intermingle depends on whether or not they are composed of the same basic "stuff." If each consists of the *same* "stuff" then after the interaction (measurement) the joint system of observer and object being measured can be described by a wave function (*quantum superposition*).

A simple way to view this is to picture combining Coke and Pepsi. Since both are made of the same stuff (soda), they readily blend together. In like manner if the observer and object being measured are each composed of the same basic stuff (matter and its motion), then their wave functions (like the Coke and Pepsi) blend together into *one wave function* that describes the joint system.

However Wigner showed that this does not occur. He demonstrated that the joint system of observer and object being measured cannot be described by one wave function. Instead the proper quantum mechanical description of their state is a mixture — akin to the state of affairs one would see trying to combine oil and water.

Properly understood, this result (quantum mixing) is the kiss of death to materialism, because it shows that unlike the 'quantum object' being measured, the observer is not composed of matter and its motion, but instead consists of a 'substance' that is non-physical, a 'substance' that is therefore not subject to quantum laws. This non-physical entity is commonly referred to in quantum physics as a "classical object."

CHAPTER 9

Materialism and Chance

Table 1 (page 79) lists the information content of several inorganic and organic systems. The table shows that a human cell is the *most complex known object in the entire universe*. [1] Although roughly 30 thousandths of an inch in diameter, it harbors a vastly complex blueprint which stretches seven feet end-to-end.

This blueprint instructs the assembly of a human brain, heart, intestines, kidneys, liver, ovaries/testes, pancreas, spleen and stomach — to name but ten, and all within only three weeks of conception! A more complete description of human body development is found in **Appendix 2** (page 85). The blueprint's instructions are expressed through a double helical structure called DNA. [2] Cell division requires the double helix to unwind at three hundred turns per second, and to reassemble new strands at ten times this speed!

Find the source of this blueprint and you will have found the origin of life. Theism asserts its source is God (Intelligence). Materialism claims it's the product of chance processes (matter and its motion). Either way we lack experimental evidence to prove God is the Source of the blueprint, or that proves matter and its motion is its source. However we *do* have experimental evidence that lead us to conclude that life's blueprint could not have originated from chance processes attending only matter and its motion.

The blueprint in question is complex — more so than any other structure in the entire universe. But what does this mean? It means that its source — whatever or Whoever that may be, had to possess an amount of informational capacity comparable to the magnitude of information that we find in the DNA blueprint of human life.

Materialism requires chance processes to have been active on matter and its motion, and to have transferred such information into a first living cell. This is not only an empty unsubstantiated claim, but one that is repudiated by information theory. [3] The certainty (inverse probability) that this could *not* have occurred by chance is so vast, as to strain the bounds of credulity. This can be illustrated as follows:

A typical 8½ by 11 sheet of paper can hold about ten thousand characters (numbers or letters). The certainty (inverse probability) that chance processes are *not* the source of life's blueprint is a number equal to a character count filling both sides of tightly stacked paper stuffed in a volume of space equal to *forty thousand universes, each 13 billion light-years wide*! [4]

By way of comparison, the certainty that we know the inverse square law of gravity is a number equal to a similar character count filling both sides of tightly stacked paper stuffed in the volume of a cube under 2 miles on a side. [5]

It's been argued that given enough time, anything is possible. Why, for example, is it not possible for chance processes to create the blueprint of the first cell? This 'isn't it possible' scenario is at best naïve, and at worst really dumb.

It's akin to asking, "Isn't it possible for a blind-folded monkey to randomly press typewriter keys to produce Einstein's Special and General Theories of Relativity (an event more likely than creating a living cell)? The answer is no, The monkey will die long before the sought after event occurs, And if we employ an eternally existing monkey, then the typewriter will wear out long before the desired manuscript appears.

These 'Isn't it possible' scenarios are at best naive, in part because they disregard the *time* required to assemble and reassemble the statistical combinations. Our universe is simply too young and small (albeit 13 billion years and twice as many light years across) to have ever spawned the magnitude of complexity required for life's blueprint.

Let's look at this from a different perspective. Many scientists have quantified the complexity of our universe. We all find that it's under 350 bits. [6] A thermodynamically closed system anemic in information is incapable of spawning a system that is robust in information. This statement is mathematically quantified by the First Law of Thermophysics, previously called the New Generalized 2nd Law of Thermodynamics. [7]

But what do these terms mean? Simply put, a 'thermodynamically closed system' can be viewed as isolated in the sense that energy neither flows in nor out. Think of a thermos filled with coffee. Hopefully the heat inside the thermos stays there and does not flow out so that the coffee stays hot. Our thermos is "thermodynamically closed" or isolated.

Now if (as most scientists believe) our universe exploded into existence from a 'singularity' (point), [8] then as it expands, it is surrounded by an outer 4-dimensional boundary across which no energy passes, thereby sustaining thermodynamic closure.

Like our thermos bottle, energy passes neither in nor out of its 4-dimensional content. [9]

Since our universe is thermodynamically closed, the First Law of Thermophysics applies. It teaches that, *on average*, information within our universe cannot increase. [10] But if this be so, how does one account for the rise in complexity of life's blueprint over time? The fossil record in **Table 2** (page 80) shows the systematic increase in complexity over several hundred million years. How can chance processes account for this rise of information in light of the First Law of Thermophysics? The answer is, 'It cannot — and here's the reason why.

Consider a collection of moving objects distributed over space. The energy exchange created by observing them alters their location and speed, creating a new distribution. Each new observation creates another new distribution. The First Law of Thermophysics requires that, on average, the information we need to describe these distributions diminishes with successive observations. In other words, matter reconfigures itself to present the least amount of information to the observer.

In effect, the First Law of Thermophysics quantifies the loss of information that attends our description of the new locations and speeds subject to predictions made from earlier measurements. This loss of information means that the distribution or structure, on average, becomes *less* complex as time passes. But the fossil record shows biological structures systematically becoming *more* complex, in violation of the First Law. Therefore something other than natural processes is at play.

Biological structures result, in part, from the assembly of amino acid residues under instructions of the DNA blueprints. The more complex the structure, the greater is the information needed to describe the blueprint. **Table 2** (page 80) outlines the fossil record.

The systematic rise in complexity with time evident in **Table 2** (page 80) means that the information contained in the successive DNA blueprints, on average, increased with time — a rise that is disallowed by the First Law within a thermodynamically closed universe. Stated differently, chance processes cannot produce information on a regular basis. But if matter and its motion cannot systematically create information, then what can? The answer is, Intelligence!

Chapter 10

The Crucial Role of the "Observer"

Quantum Mechanics has changed the way we view our world. Among other things it has given us insight into the underlying nature of physical reality. Yet be that as it may, we have had insight for decades — and without the benefit of Quantum Mechanics, into the makeup of physical matter: Its inner structure consists of tiny objects called 'molecules.'

But before we proceed — reader be warned: This is a highly *technical* chapter. In particular the nine paragraphs which follow show that our entire physical world consists of, and is constructed from *particles* — in fact our world is constructed from *hundreds of particles*. These nine paragraphs underscore the extent to which this is true. So please stay with me; anything that is worthwhile requires effort.

Let's begin with a glass of water. We know that the water is composed of countless millions of identical tiny ensembles called 'water molecules, and that each one of these structures in turn consists of still smaller assemblies called 'atoms.'

In the case of water, each water molecule contains two atoms of Hydrogen, and one atom of Oxygen — commonly referred to as "H_2O". We can picture an oxygen atom as a large white marble, and a Hydrogen atom as a small red marble. And as you might suspect, each of these atoms in turn is composed of even smaller objects called electrons, protons and neutrons. Each Hydrogen atom contains only one proton and one electron. But each Oxygen atom not only has electrons and protons (eight of each); it also contains eight neutrons.

Each electron possesses a negative unit of electric charge, while each proton has a positive unit of electric charge. These 'plus' and 'minus' charges can be viewed as two people pulling at opposite ends of a rope. If either lets go, the rope moves toward the other. But when both are pulling, the rope stays balanced between them.

Plus and minus charges likewise pull at each other. Each unit of minus charge cancels the pull of each unit of plus charge and vice-versa. Thus each atom has an equal number of electrons and protons to stay electrically balanced. But since neutrons do not have electrical charge, any number of them can exist within an atom without altering its electrical balance.

For example, although the number of neutrons in an Oxygen atom happens to equal its number of protons, a Uranium atom is found to have ninety-two protons, but one hundred forty-six neutrons. As it turns out, Hydrogen is the only atom (element) without neutrons.

By now the reader should be somewhat convinced that at least to some level scientists were able to achieve a degree of insight into the underlying nature of physical matter without the benefit of Quantum Mechanics. But the advent and application of Quantum Mechanics extended our insight into the very structure of the protons and neutrons themselves.

Quantum Mechanics enabled us to uncover the existence of many other particles beside the electron, proton and neutron. The list includes positrons, photons, weak bosons, gluons, neutrinos, mesons and baryons to name but seven. But the complete list has well over one-hundred entries — so many in fact that it has been likened to a 'particle zoo.' One particle of particular interest within this 'zoo' is the "quark."

Quarks were independently conceived by George Zweig and Murray Gell-Mann — each of whom published their work in 1964. [1], [2] Now, and well over forty years later, we know that quarks come in three distinct pairs (six quarks total). One quark of each pair has plus two-thirds of a unit electric charge, while the other has minus one-third. Quarks combine to form new particles called 'hadrons.' These hadrons come in two kinds: baryons and mesons.

Baryons arise when three quarks combine, whereas mesons are formed when a quark and its 'anti-quark' combine. An 'anti-quark' is a quark of opposite electric charge. **Table 3** (page 82) lists the names of each quark, illustrates the anti-quarks, and shows how three quarks combine to yield two kinds of baryons: a proton with plus one unit of charge, and a neutron with zero net charge (this completes the nine paragraphs).

But why have we gone through this discussion? Why have we bothered to talk about electrons and protons and neutrons and quarks? The reason is to demonstrate to you the reader — and in a rather *definitive* way that our entire physical world consists of, and is constructed from many hundreds of particles. The entire creation — from the hydrogen found after inception of the Big Bang, to the structure of a blade of grass billions of years later — is all composed of particles!

Why is this so important? It is crucial —and here's the reason why: The wave functions of Quantum Mechanics teach that physical matter consists of waves — *not particles*! These waves are smeared out over large regions of space. They become particles only when impacted through the energy exchange that occurs at the moment of observation. The act of observing (measurement) brings the particle into being.

Modern science (Quantum Physics) tells us that, in the absence of an observer (intelligence), particles really don't exist —at least not as particles. Instead all 'particles' actually exist in the form of smeared-out energy described by the wave functions cited earlier. In your mind's eye picture thousands of tiny 'dots' of ink smeared across a white sheet of paper. We can think of each dot of ink as a 'quantum' of energy, with thousands of them smeared across space before the particle comes into existence.

A particle comes into existence at the time an observer measures it. It results from the energy exchange that attends all measurements. In the language of quantum physics, the act of an observer making an observation is said to instantaneously "collapse" the particle's wave function into the particle. This is conceptually equivalent to instantaneously collapsing all of the dots into an ever-so-tiny region of space by crunching the paper into an extremely small tiny ball. This 'collapse' gives rise to the particle.

Since collapse of the wave function occurs from the impact of the energy exchanged in a measurement, and since any observation of any particle requires an act of intelligence, it follows that in the absence of intelligence, there can be no observation of particles. Stated differently, the existence of a particle requires the a priori existence of an observer to collapse the wave function to create the particle. In other words intelligence must exist before particles do, because intelligence is what creates and observes them.

The wave function collapses only when a measurement occurs. But since creation consists of vast numbers of particles, it implies that Intelligence existed before Creation i.e. that Intelligence created our physical world by collapsing wave functions into the particles that comprise all that we see and measure. These are the particles that attended the 'big-bang,' and that comprise the physical matter of our universe. Numerically, their number approximately equals the number '10' multiplied by itself 80 times.

In the light of Quantum Mechanics, how can materialism be true? We know that physical matter is composed of particles. But how can particles exist if all that exists is physical matter and its motion? Particles do not exist apart from observation, because the wave functions of Quantum Mechanics show that they instead are in the form of energy that is smeared out over space — like the dots of ink smeared across a sheet of paper. They exist as particles only when created by the collapse of their wave functions.

Considering the knowledge that is now available to us, and how none of the paths we have examined support the doctrine of materialism, a reasonable question to ask is, 'What is the basis for believing that all that exists is matter and its motion? Faith without evidence can be a dangerous thing. If ever there was a belief without rational support, materialism would win the prize for its self-serving nonsensical din.

CHAPTER 11

A Transitional Overview

This chapter summarizes scientific reasons why belief in materialism is misguided. These reasons are presented in my earlier chapters. This chapter concludes by introducing an experiment with far-reaching implications for materialism's demise, and for mankind's origin, existence and destiny. Here then is a short synopsis of the materialistic shortfall.

We said earlier that about seven billion persons are alive on earth today. But each of these is an 'intelligence'. What conceivable basis exists to believe that these arose from nonliving dust? We know that life begets life. Why then is it not more reasonable to suppose via Occam's Razor that an Intelligence populated the earth with intelligences?

We also observed that physical matter interacts with all other matter throughout the entire universe. Since such interaction appears to be absent in the private experience of human beings everywhere on earth, it suggests that human beings are more than just physical matter. If it were otherwise, how could matter interact over the space of billions of light-years and yet show no sign of doing so between two human brains just inches apart?

Is it sensible to believe that nonliving specs became living people? Is it reasonable to suppose that dead dust created live music? Is it sane to presume that inert grime came to eat and enjoy pizza — or that lifeless grunge eventually came to throw large balls down bowling alleys?

Taken at face value, these things are absurd. How then did such beliefs arise? Certainly not from anyone seeing dead dust change into people! No. They arose because large numbers of people refused to believe that theism is true. We want to be king of the mountain. We want to be the highest authority! Theism interrupts all of that.

If materialism were true, then the optical sensation we experience as 'red' would be a physical property of matter that should, could and would be describable. One reason we cannot describe the sensation humans experience as "red" may be that it resides in a realm outside physical reality. Be it color or love, music or art, pleasure or pain — these and other experiences ostensibly dwell in a realm that is inaccessible to nonliving objects.

Another more formidable conceptual obstacle to materialism is the universal observation that physical matter obeys natural laws initialized by conditions that are freely chosen by an observer. If I consist only of physical matter, how is it possible for me to impose on physical matter — initial conditions that are not a property of the physical matter of which I am solely composed? How can I as a living agent consist only of matter and its motion, and yet create initial conditions that are not a property of matter and its motion?

Our capacity to freely 'choose' among alternatives is a major obstacle to materialism. If human beings only consist of matter and its motion, then the laws that govern physical matter ought to also govern human beings. But if this be so, how can humans possess the liberty to choose freely? How can we be free to override universal laws that govern us? Materialism's sole response to this obstacle is to deny the reality of human experience!

However materialism's 'kiss of death' came when Wigner showed that the joint system of an observer and the object being measured are not described by a single wave function i.e. they do not superimpose in a manner akin to combining Pepsi and Coke. Instead the proper quantum mechanical description of their state is a mixture — akin to the state of affairs one would see trying to combine oil and water.

This means that unlike the 'quantum object' being measured, the observer belongs to a realm apart from matter and its motion — a domain of something that is non-physical, a 'substance' that is therefore not subject to quantum laws. This non-physical entity is commonly referred to in quantum physics as a "classical object" in that it is *exempt* from quantum laws.

Biological complexity presents yet another barrier to materialism. Calculations from information theory demonstrate that the certainty life could not have originated from chance processes is so vast, as to strain the bounds of credulity. Beyond this, the fossil record compounds the problem for materialism. It shows that the complexity of fossils systematically rises with time.

Biological structures result, in part, from the assembly of amino acid residues under instruction from the DNA blueprint. The more complex the structure, the greater is the information needed to describe the blueprint. Since chance processes cannot produce information on a regular basis (forbidden by the First Law), matter and its motion cannot systematically create information i.e. on average, the information cannot rise.

If all of this isn't enough, consider how particles are created. A particle comes into existence at the time an observer measures it. The transition is initiated by the energy exchange that attends the measurement. In the language of quantum physics, the act of an observer making an observation is said to instantaneously "collapse" the

particle's wave function into the particle. But how then can materialism be true?

We know that physical matter is composed of particles. But how can these particles come into existence if all that exists is physical matter (wave functions)? In the absence of observation (intelligence), what brings particles into existence? In the words of Heisenberg the properties of objects do not exist unless they are observed. [1] As Eugene Wigner has so aptly observed: The collapse of the wave function is due to the interaction of human consciousness (the observer) with the physical system (physical matter). [2]

These observations belong to the realm of science— a deep collective inquiry that covers centuries of human thought. Science has a PHD. The 'P' is for 'Predictions,' the 'H' for 'Hypothesis' and the 'D' for 'Data.' But the predictions must be logical, the hypothesis falsifiable, and the data reproducible. We make observations (gather data), interpret what we see (hypothesis) and test its predications with new observations. This new data in turn leads to a new hypothesis which makes new predictions, and so the process goes.

Reproducible events lend themselves to *scientific* inquiry whereas unpredictable events lend themselves to *statistical* inquiry. The events that ultimately led to the advent of quantum physics were reproducible. The inquiry was therefore scientific.

What of singular events? These lend themselves to *legal* inquiry. Did John Jones strike Joe Doe last Friday? Did Jane promise Jill to go up the hill on Saturday? Did God create the universe with a Big Bang? These and other such events lend themselves to legal inquiry. Here science can (and does) gather evidence in support of a particular hypothesis, but it has no proper jurisdiction in matters of origin or destiny.

Since history is a chronological record of prior singular events, its study is a legal inquiry into the past. The trustworthiness of such prior events rests upon the strength of the evidence that attends them, the dependability of the witnesses that surround them, and the reliability of the manuscripts that record them. Historical events can therefore never yield scientific conclusions. The same is true for unpredictable events. If we are to achieve scientific outcomes, the events *must* be reproducible.

CHAPTER 12

The Human Soul

An abundance of scientific observations shows that physical reality consists of something beyond matter and its motion. Nowhere is this contrast starker than between the classical and quantum objects of quantum physics — between the gap separating an observer and the reality observed. Yet if this be so, then detecting and even measuring this 'something else' beyond matter and its motion ought to be a worthwhile if not an achievable goal.

Moreover we are provided with a hint of where to look for this "something else" in the quantum dichotomy of an "observer" (the 'something else') and the "reality observed" (matter and its motion). To use terms perhaps more familiar to the reader, we will borrow from the parlance of religious culture to identify an "observer" with 'spirit,' and the "reality observed" with a human body. Although the former can be viewed as semantics, the latter is another matter.

All observations of physical reality (matter and its motion) occur under an energy exchange in which some form of external energy stimulates sensory structures within the body. Thus an observer's body participates in all measurements of physical reality. The measurement process is triggered when neurological pathways channel external energy into interpretive brain centers. These in turn form 'patterns of meaning' within the brain.

Stated differently, the world we inhabit is *determined* by the body we possess. Acquire a new body with a different design and you will reside in and perceive a new world. This is not because you are your body, but instead because the reality we perceive is established by the energy patterns we observe — patterns defined by our sensory structures and brain.

But if we are *not* our body, then 'who' or 'what' are we? In the language of quantum physics we are the 'observer.' The Bible frames the picture this way: It declares we are the union of 'spirit' and 'body'. [1] This is not unlike the dichotomy of the 'observer' and 'physical matter.' Theological tradition identifies the union of spirit and body as a 'soul' thereby viewing a 'soul' and a 'person' as synonymous.

However as commonly used the term 'soul' denotes one's 'spirit' per se. Thus the word 'soul' can be used to either mean a 'person' or a person's spirit. But irrespective of whether we regard the soul as one's spirit, or as the entire person (body and spirit) we are nonetheless confronted with the existence of 'something else' beyond matter and its motion. If so, then we are faced with the formidable challenge to detect, measure and uniquely quantify the substance of this 'something else' regardless of its genre or venue.

How shall we go about doing this? It's at this juncture we observe that all of the examples thus far cited to refute materialism share a common denominator. To illustrate, consider the following list:

Concealment	Consciousness	Color
Initial Conditions	Classical Object	Observer
Complexity	Information	Measurement

Each of these nine concepts were discussed in preceding chapters, and applied in a way that refutes materialism. In each case these (and other) observations were seen to display inconsistencies if — all that exists is matter and its motion. The significant thing to note however is that in each of the instances cited in the preceding chapters the contradiction is resolved by introducing the existence of an observer (spirit or soul) i.e. by the existence of a living human consciousness.

If this be so, then it means that every living human body should contain a spirit or soul — what in quantum physics is referred to as an 'observer.' We can therefore conceptually distinguish between a human body in the form of matter and its motion, and a spirit or soul in the form of an albeit invisible but not necessarily undetectable substance. The challenge which confronts us is that if such an invisible substance does exist, where does one go to find it, what does one do to detect it — and how does one quantify or measure it?

We can approach answering such questions by looking at some of the outcomes implied by the concepts listed above. For example, consider human 'consciousness.' Somehow, some way we are living agents who are conscious of the witness we bear to the data of our experience. This means there exists a coupling — albeit ill-defined, but nonetheless a linkage between us as living agents who collect data, and the physical reality that we see and measure. This implies that a component exists *which is common* to the physical world that we do see, and the observer or spirit or soul which we do not see.

Likewise consider 'color.' We know we can see it. But where does it reside? I do not mean the electromagnetic wave giving rise to the perceived sensation — but the sensation itself. If what we perceive belonged to the space-time continuum we would be able to describe it. But we know the sensation we perceive resides elsewhere because it evades description. Tell a person blind from birth what you see when you see the color 'red.' You can't! It's not possible. But yet an observer 'sees' it, again showing a linkage between the observer or spirit or soul, and the physical reality which gives rise to the perceived sensation.

Perhaps there is no more striking example of this linkage than a measurement that spawns the instantaneous collapse of a wave function. Prior to collapse we have a smeared out wave with tentacles throughout the vast expanse of space and time. After collapse the wave disappears, having coalesced into an infinitesimally small localized particle. The transformation is catalyzed by the energy exchanged in the measurement. After collapse the wave is perceived to be a particle by the observer.

However whereas the wave and particle belong to realm of physical reality, this is not so of the observer. S/he is a classical object and therefore immune to quantum laws. Nonetheless the living agent is able to observe the post-transformation existence of the particle, thereby disclosing the existence of linkage of some kind between the observer or spirit or soul, and the physical realm embraced by the measurement of the particle. Yet how is the observer able to observe the particle?

Although we are unable to answer this question, we can extract an important clue that would appear to attend the nature of this linkage. Given that an observer is able to observe the existence of the particle, and given that the particle resides in the realm of physical reality, we surmise that at least one component of an observer must likewise reside in the realm of physical reality. Moreover since physical reality is largely composed of space-time, it strongly suggests that like the particle, *a gravitational component must attend the observer*. Otherwise the absolute disunion would render an observation impossible.

We can discern further insight into the linkage between an observer and the physical realm measured by considering the following thought experiment: A message is displayed, instructing me to raise my right arm. I do so. The message is then changed, instructing me to raise my left arm. I do so. Next the message is randomly changed every ten seconds, instructing me to either raise my right arm or my left arm. I comply for some period of time and then quit.

Let's now revisit this thought experiment where I was first instructed to raise my right arm. Let's ask, "Why did my right arm lift but not my left arm?" The simple answer is that electrons traveled from my brain, and along neurological wires into muscle tissue in my right arm, thereby causing certain muscles in my right arm to contract. My left arm failed to move because the corresponding electrical current was absent. Likewise when I was instructed to raise my left arm, its muscles contracted due to similar electron flow from my brain along neurological wires in my left arm. My right arm failed to move at that time due to the absence of such electrons in the muscle tissue of my right arm.

However the more interesting question to ask is, "What caused electrons to travel from my brain and along neurological wires to orchestrate biological events that eventually raised my right versus my left arm?" Who or what causes electrons to be launched within my brain? Once the electron flow is initiated, who or what defines the path of their travel? One answer to such questions is that the choices I make as a living agent act to collapse wave functions within my brain, thereby creating particles (electrons) which then flow along neurological paths identified by the location of their origination.

Although the wave function itself is smeared out over space and time, its collapse occurs in a well-defined region of space which ultimately creates a localized event in space-time. This view is self-consistent with the 'observer' of quantum physics and yields additional insight into the 'linkage' between an observer (spirit or soul), and the physical realm embraced by the existence of the wave function and the particle. The important thing to note is that the wave function is seen to be the means by which the observer acquires access to the physical realm of space-time.

Since the wave function is described by variables that include space, time and mass, it underscores our earlier conclusion that a gravitational component attends an observer. This understanding was suspected more than one hundred years ago by a physician named Duncan MacDougal — at a time when quantum physics didn't exist. The next chapter discusses how this came to be, and the unusual experiment he devised to probe the existence of the human soul.

Chapter 13

Weighing the Soul

Duncan MacDougall was a reputable physician who resided in Haverhill, Massachusetts. He had held for many years his growing conviction that if the spirit of a person continues to exist after death then the individual must exit the physical realm as a 'space occupying body'. Stated differently, Dr. MacDougall believed that the psychic functions continue to exist as a separate personality after the death of the brain and body, Although he accepted other views, he believed it unthinkable that a conscious personality could continue to exist and have being with ongoing personal identity and yet not occupy space.

MacDougall therefore believed that the continuance of conscious life and personal identity after death implied (if not required) the existence of a space-occupying substance of some kind. This conviction led him to ask the question, "Does this substance have weight?" In the words of MacDougall, "Is it ponderable?" (meaning, 'Is it capable of being weighed?'). This question ultimately drove MacDougall to devise an experiment with far-reaching implications — measurements that would impact human religious systems, and whose outcome would impose formidable constraints on the very destiny of mankind.

In the words of MacDougall [1]:

"I am aware that a large number of experiments would require to be made before the matter can be proved beyond any possibility of error, but if further and sufficient experimentation proves that there is a loss of substance occurring at death and not accounted for by known channels of loss, the establishment of such a truth cannot fail to be of the utmost importance."

MacDougall reasoned that if a ponderable space-occupying substance of some kind did exist, and if it enveloped the conscious life and personal identity of a human being, then when that person died, a weight loss should occur due to the absence of this substance (conscious life) from the dead body. To this end MacDougall created a beam balance that proved accurate to two-tenths of an once (5.7 grams). [2]

He placed a bed upon a light framework which had been built with great care upon very delicately balanced platform beam scales. When placed at balance, one-tenth of an ounce would lift the beam up close to the upper limiting bar. Another one-tenth ounce would bring the beam up and keep it in direct contact with the limiting bar. Then if the two-tenth weight was removed, the beam would drop to the lower bar and then slowly oscillate until balance was reached again.

Fluid loss was continually measured during the several hours which typically preceded death. This fluid weight loss was very slow and due to evaporation of moisture from the nasopharyngeal and bronchopulmonary and buccal mucous membrane that accompanied respiration, as well as evaporation of moisture from cutaneous perspiration (sweat).

Although bowel movements would be unusual, if one occurred it would have remained on the bed except for a slow evaporation of feces' fluid. When urination occurred, the bladder evacuated at most about one-quarter of an ounce of urine; this would remain on the bed and then evaporate.

Respiration of all but residual air in the lungs was separately measured by two physicians (Drs. MacDougall and Sproull). It was found to have no effect. The rate of the weight loss from evaporation when all fluids were taken into account was found to lie in the range: 0.75 - 1.00 ounces per hour.

MacDougall accepted persons for his experiment who were on the brink of death, and as circumstances permitted. Those suffering with tuberculosis were best for his purposes because the disease produces great exhaustion with little or no muscular movement, thereby enabling the beam to be better balanced.

In the several hours prior to death, the end of the beam was maintained slightly above balance, near the upper limiting bar. This was to bias against a positive result. When death occurred, the end of the beam abruptly dropped with an audible stroke, and hit against the lower limiting bar where it remained without rebound at the time of death. It is important to note, however, that the slow gradual evaporation of fluids cannot account for the sudden drop in weight observed at the moment of death.

Dr. MacDougal measured the change in weight of a total of six persons at the time of their deaths. However he believed that he was justified in recording the abrupt weight loss at the time of death of only four of these six persons. His reasons for omitting two of his measurements, along with his assessment of all six weight loss experiments were published in American Medicine [1] and in the Journal of the American Society for Psychical Research [2]. Dr. MacDougall's assessment of these six measurements is summarized in **Appendix 3** (page 87); an overview of the results obtained is given in **Appendix 4** (page 91).

The average of all weight loss measurements at death (four persons) is about 0.75 ounces (21 grams). As the reader can well imagine, the implications of an abrupt weight loss at death is staggering. If true, it implies (confirms ?) the continuance of the conscious life and personal identity of a human being beyond the grave. However such continuance goes against our nature in that it implies the existence of an authority higher than our self. We are inclined therefore to dismiss MacDougall's results as a curiosity at best. Stated differently, we will not have that seem true which contradicts our passions and our affections.

How then can we dismiss the sudden weight loss at death which MacDougall and his coworkers observed? Some have claimed that his results could not be reproduced. This seems a nonsensical claim in that I am unaware of anyone who has published even one attempt to repeat what he did. Others have questioned MacDougall's methodology saying it was suspect. This too is a bogus allegation. His methodology leaves us little wiggle room in that he experimentally demonstrated that an abrupt weight loss occurs when a human being dies, and that this is consistent with a ponderable substance leaving the body at death.

Still others allege that his ability to measure changes in weight were imprecise. To the contrary, the rate of weight loss due to evaporation from all fluids was about one ounce per hour, whereas the average rate of abrupt weight loss at death was about 0.75 ounces per two seconds, corresponding to well over 1000 ounces per hour. Therefore the allegation is without merit, especially when we consider the 0.2 ounce sensitivity of his beam balance.

There is also the claim that MacDougall's sample size was too small. All would agree that greater numbers of people on the brink of expiring and available to be weighed at death would have given greater credence to the experimental outcome. Nonetheless absence of greater numbers of people does not negate the validity of the measurements taken. It is noteworthy that even the two cases dismissed still showed what the other four cases showed: an abrupt weight loss of the order of an ounce.

When nuclear technology was developed,, and Fermi's theory of neutron capture needed to be verified within processes required to produce the first atomic bomb, no one complained that we only had a sample size of one. Just one explosion was all that we needed to verify his results and to later manufacture atomic bombs and to drop them on Japan. In MacDougall's case we have at least four confirmations (if not six).

Irrespective of any other consideration, Dr. MacDougall's measurements conducted on at least four human beings show that an abrupt weight reduction occurs at death which is not accounted for by known channels of loss.

Chapter 14

Do Animals Have Souls?

An interesting question which was pursued by Dr. MacDougall is, "Do animals have souls?" In one instance he set about to answer this question by measuring the weight change of fifteen dogs at death. These dogs ranged in weight from fifteen to seventy pounds. [1]

Two drugs were administered to secure the calm, peace and quiet needed for the measurements. MacDougall noted that he would have preferred to use dogs that were dying of some illness that rendered them exhausted and free from struggle, but that it was not his good fortune to obtain such dogs.

Unlike the 0.2 ounce sensitivity of the beam balance used to measure the weight loss at time of death of humans, here the scales with the total weight upon them were sensitive to one-sixteenth of an ounce. What were the results of measuring the weight change of dogs at their time of death? Quoting from Dr. MacDougall's own words [1]:

"The same experiments were carried out on fifteen dogs, surrounded by every precaution to obtain accuracy and the results were uniformly negative, no loss of weight at death."

This is a remarkable result. Whereas the scales used to measure the weight change in dogs at the time of death were over three times more sensitive than those used to measure the weight change in humans, nonetheless an abrupt weight loss was observed at death in all six humans measured while no weight loss was observed in any of the fifteen dogs. In the words of MacDougall [2]:

"If it is definitely proved that there is in the human being a loss of substance at death not accounted for by known channels of loss, and that such loss of substance does not occur in the dog as my experiments would seem to show, then we have here a physiological difference between the human and the canine at least and probably between the human and all other forms of animal life."

Dr. MacDougall's assessment of the results of his measurements as each animal expired is given in **Appendix 5** (page 92).

This distinction between humans and animals is reminiscent of a passage found in the Book of Genesis chapter One, verse 26 of the Holy Bible which reads,

"Then God said, "Let Us make man in Our image, according to our likeness;"

If man is created in the image of his Creator, would we not expect man to be a moral being possessing a spirit or a soul? Since animals are not moral creatures we would not expect them to possess a spirit or a soul. This is consistent with MacDougall's results: an abrupt weight loss at time of death for humans but not for dogs.

What's of particular interest is that the Bible prognosticates MacDougal's experimental outcome. The Bible teaches that animals do not possess a soul. In order to understand where and how the Bible teaches this we turn to the eighteenth book of the Old Testament — to the Book of Job. The opening chapter introduces Job as a man who was "blameless, upright, fearing God and turning away from evil." Job was also a wealthy man. The first chapter (verse 3) tells us he possessed 500 female donkeys, 500 oxen, 3000 camels and 7000 sheep.

Job also had seven sons and three daughters. Each son had a house and held a feast on 'his day', inviting his six brothers and three sisters to also eat and drink. When the days of feasting completed their seven day cycle, Job would convene and consecrate them by means of burnt offerings in the early morning. His concern that his sons might have sinned and cursed God in their hearts prompted Job to do this regularly.

The first chapter of the book of Job goes on to say that one day Job's adversary (Satan) came among the sons of God when they were presenting themselves before the Lord. Upon the Lord telling Satan that there is no one like Job on the earth, a 'blameless and upright man, fearing God and turning away from evil,' Satan replies, and credits Job's good fortune to the Lord's protecting and blessing him.

Satan then says that if the Lord will remove all of Job's possessions, Job will curse the Lord to His face. So the Lord transfers all that Job possesses over to Satan's power. Then on the day Job's sons and daughters were eating and drinking wine in the oldest brother's house the Sabeans attacked the oxen as they were plowing and the donkeys as they were feeding beside them. They took the animals and also killed most all of Job's servants who were attending them.

Then fire from heaven burned and consumed Job's sheep and the servants attending them. Next the Chaldeans raided and took Job's camels and killed the servants attending them. Then the oldest brother's house was struck by a great wind and fell upon and killed all of Job's children. Through all of this Job did not sin nor did he blame God.

The second chapter (verse 1) tells us that Satan and the sons of God came again to present themselves before the Lord. Here the Lord tells Satan that Job "still holds fast to his integrity" even though Satan incited the Lord against Job, to ruin him without cause. Satan replies that if the Lord will harm Job's bone and flesh, then Job will curse the Lord to His face. The Lord then gives Job over to Satan, instructing Satan to spare Job's life.

Verse 7 tells us that upon departing from the Lord's presence, Satan "smote Job with sore boils from the sole of his foot to the crown of his head. When Job's three friends (Eliphaz, Bildad, Zophar) learned of Job's adversity they made an appointment together to sympathize and comfort Job. The closing verse of Chapter 2 tells us:

> They sat down on the ground with Job for seven days and seven nights with no one speaking a word to him, for they saw that his pain was very great.

The next thirty-nine chapters record dialogues between Job, his three friends and God as to how and why Job has been made to suffer this way. The last chapter forty-two shows Job with humility of heart confessing to God,

> "I have heard of Thee by the hearing of the ear, but now my eye sees Thee; therefore I retract, and I repent in dust and ashes."

In other words, up to this time Job had only heard about God, but now the suffering he had been made to endure enabled Job to see God, in the sense of Job now having access to God through a close, intimate personal relationship.

We are told (verse 10) that the Lord restored the fortunes of Job, *increasing all that Job had twofold.*

Job once had 500 female donkeys; he now is given 1000.

Job had 500 oxen; he is now given 1000.

Job once had 3000 camels; the Lord now gives him 6000.

Job had 7000 sheep; he is now given 14,000 sheep.

God *doubled* all that Job once had.

Job had seven sons and three daughters; the Lord now gives Job seven sons and three daughters.

Please re-read this carefully.

Whereas God doubled all of Job's animals, he only had to *replace* Job's children in order to double them!

If human beings have a soul that continues on after death, then replacing a person after his or her death yields two souls — one who continues on after death, and the one who is his or her replacement.

Thus the Bible is teaching here that since the animals had to double in number in order to increase twofold, they do not possess a soul. However this is not true of human beings because for these to increase twofold, they only needed to be replaced in equal number i.e. seven sons and three daughters (not fourteen and six).

I recall giving a series of lectures in 1986 in the Seattle, Washington area in which I mentioned the MacDougall experiment. Afterwards I was approached by a wealthy man who had lost his wife about two years earlier and who offered me a large sum of money to repeat MacDougall's experiment. During the course of assessing his proposal two books emerged as necessary to an objective assessment.

The first was a 221 page loose-leaf binder by Bernard D. Reams Jrs. & Carol J. Gray entitled Human Experimentation. This pertained to Federal Laws, Legislative Histories, Regulations and Related Documents. The other was a 467 page Annotated Resource Guide by Hannelore-Wass et.al. entitled Death Education II. This was a 'Series in Death Education, Aging and Health. Several meetings were scheduled June 1 - 5 to evaluate the proposal — which had been dubbed the 'MacDougall Project.'

It was at the last of these meetings that I declined the proposal to repeat MacDougall's experiment. I did so for several reasons, one of which was because people were no longer dying of tuberculosis in 1986. This was not true 85 years earlier in 1901 when MacDougall did his experiments.

Another reason that I declined was the excessive Federal laws, State regulations and Medical Board rules and red tape prevalent in 1986 but absent in 1901. The third reason I declined to perform MacDougall's experiment was that it was regarded by a surprisingly large number of researchers as 'ghoulish.'

Although I did not share such psychological concerns, it did impact my overall assessment of the value of the project; and while it carried the least weight of all the factors considered in making the final decision to abort the project, it nonetheless did contribute to the my decision to decline the offer. To my knowledge no one has attempted to duplicate MacDougall's experiment.

CHAPTER 15

The Near Death Experience

MacDougall's conclusion that the human soul survives death was based upon measurements of a sudden loss in weight at the time of human death. As was noted earlier, to my knowledge no one has ever attempted to duplicate MacDougall's experiment. Nonetheless the outcome of this experiment viz. continuance of the human soul beyond death is an occurrence shared by growing numbers of people who report undergoing a Near Death Experience.

This Near Death Experience (NDE) is said to embrace the clinical death of a person for some short amount of time, anywhere from several minutes up to the better part of an hour, after which s/he is resuscitated back to life. Later the person relates experiences that they recall took place during the short time they were clinically dead. In effect they report what they remember seeing and hearing beyond the grave.

This is quite different from the MacDougall experiment in which the person dies and remains dead. In the NDE the person dies and then is revived back to life to tell what s/he saw while in a state of death. However both queries share the following common feature: Either inquiry supports the belief that something from the realm of our world passes through and beyond the grave and into another separate realm. One noteworthy aspect of the NDE is that a significant number of those who return from clinical death tend to report having experienced the same kinds of things beyond the grave.

The NDE begins with a sense of peace and well-being, a wonderful state of euphoria. This is followed by the sensation that you are removed and separate from your body — but yet viewing it while floating outside and apart from it. If your short term clinical death is the result of a medical procedure, a typical sensation is one of floating near the ceiling while observing medical personnel working with your body on a table below.

What next occurs is an awareness of motion, usually in or toward a tunnel with a light at its end — a light of indescribable brightness, and with whom telepathic communication occurs that overwhelms you with a sense of unconditional love and acceptance. Along the way some report seeing beings of light or relatives or friends dressed in white. The further into the tunnel you move the greater your sense of well-being and the further into the tunnel you want to go.

But you sense a point beyond which you will be unable to return. When you are told you must return you don't want to do so. Although you try to resist your body pulls you like a magnet while the tunnel acts to push you out of the tunnel and towards your body. As you return into your body it feels like you are jumping into a lake of ice-cold water.

The NDE described in the above three paragraphs can be viewed as somewhat 'typical' of those reported. However individual reports can be found that vary widely from what is reported above. There is also a class of the NDE which is unpleasant at best, and where the person encounters fiendish beings of demonic origin who express evil intent.

Yet regardless of the specific content of the NDE they all display one common feature: Each report provides us with testimony that the person appears to continue as an individual personality beyond the grave. But there's a "what if" to this conclusion. What if the brain undergoes an organic change at the time a human dies, creating electrical waves that neurological brain centers interpret as a tunnel with a bright light and a handful of relatives or friends?

Would this not explain the experiences reported by the person resuscitated to life during the time s/he was clinically dead? This kind of explanation is underscored by the find that an 'out of the body' experience can be simulated by electrically stimulating the angular gyrus in the brain [1]. This discovery ostensibly links the near death experience to the stimulation of a certain part of the brain, thereby providing a possible biological explanation for the near death experience (NDE).

The problem with this explanation, however, is that it fails to account for the person's consciousness that exists during the time s/he is clinically dead. Stated differently, if brain activity creates consciousness, how can consciousness exist in the NDE where brain activity is nonexistent? Were consciousness to result from brain activity, a near death experience would not exist But we know that it does exist, implying that human consciousness is something above and beyond brain tissue (physical matter).

Consider a NDE that results from cardiac arrest. There are numerous examples of persons who obtain medical information when they are clinically dead and, as they describe it, are floating near the ceiling watching the medical staff perform specific procedures on their body as it lies on the operating table. [2] Some of this information is on panels elevated above the operating table and can only be seen from above, and some of the procedures, such as resuscitation, are only done when the patient is clinically dead.

Light and Death (a recent book by cardiologist Michael Sabom, M.D.) documents the near death experience of a woman who had a life-threatening aneurysm in her brain. Its size and location disallowed its removal using traditional surgical techniques and she was referred to a specialist who had developed a specialized 'hypothermic cardiac arrest' procedure which required lowering the woman's body temperature to sixty degrees Fahrenheit, halting her heartbeat and breathing, and draining all the blood from her head. Her flattened brain waves confirmed that for all practical purposes she was dead.

After her aneurysm was removed, she was brought back to life. After awakening the woman proceeded to relate her near death experience, accurately describing surgical instruments, medical procedures and the details of a number of "out-of-the-body" observations she made during her surgery while floating near the ceiling.

This is one of a number of NDE cases where persons accurately convey medical details they saw and heard while out of their body, and during a time period when medical personnel confirm they were clinically dead. Since such persons are bearing witness to the data of their experience while they are in both a conscious and a clinically dead state, it implies that human consciousness continues to exist after death apart from the brain.

Table 1

Informational Bits

Physical Structure	Complexity
Typical Book	32
Public Library	48
Library of Congress	58
Human Knowledge	73
Planet Earth	160
Solar System	170
Universe	235
Simple Protein	300
Convoluted Protein	1500
Simple Bacterium	7,000,000
Human Cell	20,000,000,000

TABLE 2

The Systematic Rise in Fossil Complexity

AGE	ERA	PERIOD	DESCRIPTION
3.7Byr	**Geozoic**	Archean	oxidized Fe layers;
3.3Byr			bacteria-like forms
2.0Byr		Stromatolian	blue-green algae (cyanobacteria)
1.6Byr		Eukaryoian	green algae
600M	**Paleozoic**	Cambrian	invertebrate
500M		Ordovician	jawless fish
425M		Silurian	fish with jaws
405M		Devonian	insects & amphibians
345M		Carboniferous	reptiles
280M		Permian	mammal-like reptiles
230M	**Mesozoic**	Triassic	dinosaurs
185M		Jurassic	birds & angiosperms
135M		Cretaceous	modern insects
65M	**Cenozoic**	Tertiary	mammals
2M		Quaternary	ice age

EPOCH

65M	Paleocene	Class
58M	Eocene	Order
36M	Oligocene	Family
25M	Miocene	Genus
12M	Pliocene	Species
2M	Pleistocene	Homo sapiens
10K	Holocene	Civilizations

TABLE 3

Combining Quarks

Quark Charge	$-\frac{2}{3}$	$+\frac{2}{3}$	$-\frac{1}{3}$	$+\frac{1}{3}$	Neutrinos	Leptons	Lepton Charge
Quarks	$\bar{U}$	U	D	$\bar{D}$	Electron Neutrino	Electron	−1
	$\bar{C}$	C	S	$\bar{S}$	Muon Neutrino	Muon	−1
	$\bar{T}$	T	B	$\bar{B}$	Tau Neutrino	Tau	−1
Force Agents	Photon		Gluon		Weak Force Mediator	Weak Force Particle	± 1

Up & Down Quark Charges	$2 * (+\frac{2}{3})$	$-\frac{1}{3}$	Net Charge **+1**
2 Up & 1 Down Quarks	**2 Up Quarks**	**1 Down Quark**	**Proton (Baryon)**

Up & Down Quark Charges	$+\frac{2}{3}$	$2 * (-\frac{1}{3})$	Net Charge **0**
1 Up & 2 Down Quarks	**1 Up Quark**	**2 Down Quarks**	**Neutron (Baryon)**

Up Quark & Down Antiquark Charges	$+\frac{2}{3}$	$+\frac{1}{3}$	Net Charge **+1**
1 Up Quark & 1 Down Antiquark	**1 Up Quark**	**1 Down Antiquark**	**Plus Pion (Meson)**

APPENDIX 1

The Interaction of Physical Matter

We know that physical matter interacts throughout the universe. For example, consider how the sea rises and falls due to the gravitational pull of our moon and sun. Our sun is almost 93 million miles away, yet it interacts with our oceans.

This interaction is so strong that land is also affected. Our continents rise and fall just like the sea, twice daily due to the moon and sun pulling it. The point is, physical matter interacts everywhere with all other physical matter — and to such an extent that Coriolis forces from the gravitational interaction of stars billions of light-years distant from earth help produce the swirl in the water of a toilet on earth when it is flushed!

But if you and I consist only of physical matter, and since physical matter interacts with all other physical matter in these ways, how can we have private thoughts? How can what you see be seen by only you? How can what I hear be heard by only me? Do not all of our senses (including touch and smell) eventually reduce to electric waves traveling along and through hundreds of millions of neurons in our brain?

How can it be that physical matter interacts with all other physical matter across billions of light-years of space, and yet the physical matter of our brains not interact if I were to stand just one foot away from you?

Indeed our entire claim to a 'private experience' rests in the conviction that interactions between the physical matter in the brains of two people standing physically close to one another do not occur. Were it otherwise my conscience might be yours and vice-versa.

Here's the point: Since physical matter interacts throughout the entire universe, and since such interaction appears to be absent in the private experience of human beings everywhere on earth, it suggests that human beings are more than just physical matter.

How is it that physical matter interacts over the space of billions of light-years and yet shows no sign of doing so between two human brains that are just inches apart?

Appendix 2

Human Fetus Progression in the Womb

Human Conception

Biological knowledge demonstrates that soon after human conception billions of cells form and arrange themselves into highly organized structures under instructions from the DNA blueprint in the woman's womb. This blueprint begins to function the moment the sperm fertilizes her egg.

Two Weeks

After 2 weeks, blood vessels appear. Several days later, two of these join to form a heart which, by the end of the 3rd week starts pumping blood. Although the body in her womb at this time is less than 0.7 of an inch long, its tiny heart is already pumping blood-filled oxygen to its brain.

Four Weeks

By the 4th week a cellular boundary has formed that separates the abdomen from the back; another border marks the advent of the liver, while yet others show the start of the anulis umbilicalis, coccyx and naval. The arrival of shoulder grooves is also apparent, even though the body is now only 1.7 inches in length.

Eight Weeks

By 8 weeks these billions of cells have formed ears, eyes and lungs for future use, and a nervous system that embraces optical, olfactorius (smell), calf, elbow and hip nerves, as well as nerves along the outer soles of the feet.

The nervous system also includes the cerebellum and medulla oblongata.

Although only 8 weeks old, there are now eyelids, fingers and toes, and neurulation —the process that forms the spinal cord. All of this in a body now only 7 inches long. Development over these 8 weeks occurs in 23 "Carnegie stages" that can be seen at: http://embryology.med.unsw.edu.au/

Twelve Weeks

After 12 weeks, cell differentiation is done, and all of the organs have now formed, including the liver, pancreas, spleen, kidneys, heart, brain, stomach, intestines and ovaries/testes. The incomprehensible complexity of any one of these structures cannot be overstated.

Gray's Anatomy

Anyone who doubts this need only open the1995, 38th British Edition of Gray's Anatomy one page at a time — each a superficial description of a subsystem of the human body. Page after page reveals intricate structures that strain the bounds of credulity, but yet that are functionally designed to harmonize with all of the other organs of the body to sustain life —and it is all complete by 12 weeks! No one on earth today can explain how such structures arose from the basic properties of physical matter — I repeat — No one!

Appendix 3

Human Weight-loss Measurements

Dr, Duncan MacDougal measured the weight change of six persons at the time of their deaths. The following is Dr. MacDougall's assessment of the abrupt weight loss observed as each expired (published in American Medicine, Volume 2, No.4, April, 1907).

"My first subject was a man dying of tuberculosis. It seemed to me best to select a patient dying with a disease that produces great exhaustion, the death occurring with little or no muscular movement, because in such a case the beam could be kept more perfectly at balance and any loss occurring readily noted.

The patient was under observation for three hours and forty minutes before death, lying on a bed arranged on a light framework built upon very delicately balanced platform beam scales.

The patient's comfort was looked after in every way, although he was practically moribund when placed upon the bed. He lost weight slowly at the rate of one ounce per hour due to evaporation of moisture in respiration and evaporation of sweat.

During all three hours and forty minutes I kept the beam end slightly above balance near the upper limiting bar in order to make the test more decisive if it should come.

At the end of three hours and forty minutes he expired and suddenly coincident with death the beam end dropped with an audible stroke hitting against the lower limiting bar and remaining there with no rebound. The loss was ascertained to be three-fourths of an ounce.

This loss of weight could not be due to evaporation of respiratory moisture and sweat, because that had already been determined to go on, in his case, at the rate of one sixtieth of an ounce per minute, whereas this loss was sudden and large, three-fourths of an ounce in a few seconds.

The bowels did not move; if they had moved the weight would still have remained upon the bed except for a slow loss by the evaporation of moisture depending, of course, upon the fluidity of the feces. The bladder evacuated one or two drams of urine. This remained upon the bed and could only have influenced the weight by slow gradual evaporation and therefore in no way could account for the sudden loss.

There remained but one more channel of loss to explore, the expiration of all but the residual air in the lungs. Getting upon the bed myself, my colleague put the beam at actual balance. Inspiration and expiration of air as forcibly as possible by me had no effect upon the beam. My colleague got upon the bed and I placed the beam at balance. Forcible inspiration and expiration of air on his part had no effect. In this case we certainly have an inexplicable loss of weight of three-fourths of an ounce. Is it the soul substance? How other shall we explain it?

My second patient was a man moribund from tuberculosis. He was on the bed about four hours and fifteen minutes under observation before death. The first four hours he lost weight at the rate of three-fourths of an ounce per hour. He had much slower respiration than the first case, which accounted for the difference in loss of weight from evaporation of perspiration and respiratory moisture.

The last fifteen minutes he had ceased to breathe but his facial muscles still moved convulsively, and then, coinciding with the last movement of the facial muscles, the beam dropped. The weight lost was found to be half an ounce. Then my colleague auscultated the heart and found it stopped. I tried again and the loss was one ounce and a half and fifty grains.

In the eighteen minutes that lapsed between the time he ceased breathing until we were certain of death, there was a weight loss of one and a half ounces and fifty grains compared with a loss of three ounces during a period of four hours, during which time the ordinary channels of loss were at work. No bowel movement took place. The bladder moved but the urine remained upon the bed and could not have evaporated enough through the thick bed clothing to have influenced the result.

The beam at the end of eighteen minutes of doubt was placed again with the end in slight contact with the upper bar and watched for forty minutes but no further loss took place.

My scales were sensitive to two-tenths of an ounce. If placed at balance one-tenth of an ounce would lift the beam up close to the upper limiting bar, another one-tenth ounce would bring it up and keep it in direct contact, then if the two-tenths were removed the beam would drop to the lower bar and then slowly oscillate till balance was reached again.

This patient was of a totally different temperament from the first, his death was very gradual, so that we had great doubts from the ordinary evidence to say just what moment he died.

My third case, a man dying of tuberculosis, showed a weight of half an ounce lost, coincident with death, and an additional loss of one ounce a few minutes later.

In the fourth case, a woman dying of diabetic coma, unfortunately our scales were not finely adjusted and there was a good deal of interference by people opposed to our work, and although at death the beam sunk so that it required from three-eighths to one-half ounce to bring it back to the point preceding death, yet I regard this test as of no value.

My fifth case, a man dying of tuberculosis, showed a distinct drop in the beam requiring about three-eighths of an ounce which could not be accounted for. This occurred exactly simultaneously with death but peculiarly on bringing the beam up again with weights and later removing them, the beam did not sink back to stay for fully fifteen minutes. It was impossible to account for the three-eighths of an ounce drop, it was so sudden and distinct, the beam hitting the lower bar with as great a noise as in the first case. Our scales in the case were very sensitively balanced.

My sixth and last case was not a fair test. The patient died almost within five minutes after being placed upon the bed and died while I was adjusting the beam.

In my communication to Dr. Hodgson I note that I have said there was no loss of weight. It should have been added that there was no loss of weight that we were justified in recording.

My notes taken at the time of experiment show a loss of one and one-half ounces but in addition it should have been said the experiment was so hurried, jarring of the scales had not wholly ceased and the apparent weight loss, one and one-half ounces, might have been due to accidental shifting of the sliding weight on that beam. This could not have been true of the other tests; no one of them was done hurriedly.

My sixth case I regard as one of no value from this cause."

APPENDIX 4

Human Weight-loss Results

Patient	Cause of Death	Fluid Evaporation	Weight Drop
1 m	Tuberculosis	1.00 oz / hour	0.75 ounces
2 m	Consumption	0.75 oz / hour	0.50 then 1.60
3 m	Tuberculosis	----------	0.50 then 1.00
4 f	Diabetic Coma	----------	0.44 ounces *
5 m	Tuberculosis	----------	0.38 ounces
6 m	Tuberculosis	----------	1.50 ounces *

*	denotes measurement was discarded as having no value
----------	denotes fluid loss went unpublished
m/f	denotes male / female

Appendix 5

Canine Weight-loss Measurements

Dr, Duncan MacDougal measured the weight change of fifteen dogs at the time of their deaths. The following is Dr. MacDougall's assessment of the results of his measurements as each expired (published in American Medicine, Volume 2, No. 4, April, 1907).

♦❖♦

"The same experiments were carried out on fifteen dogs, surrounded by every precaution to obtain accuracy and the results were uniformly negative, no loss of weight at death.

A loss of weight takes place about 20 to 30 minutes after death which is due to the evaporation of the urine normally passed, and which is duplicated by evaporation of the same amount of water on the scales, every other condition being the same, e.g., temperature of the room, except the presence of the dog's body.

The dogs experimented on weighed between 15 and 70 pounds and the scales with the total weight upon them were sensitive to one-sixteenth of an ounce. The tests on dogs were vitiated by the use of two drugs administered to secure the necessary quiet and freedom from struggle so necessary to keep the beam at balance.

The ideal tests on dogs would be obtained in those dying from some disease that rendered them much exhausted and incapable of struggle. It was not my fortune to get dogs dying from such sickness."

References

Prologue

[1] Robertson A.T. *Introduction to the Textual Criticism of the New Testament* (1925) Broadman Press p.70

[2] After Jesus bodily rose from the grave He appeared to:

Peter (Cephas)	1 Corinthians 15:5
James (Jesus' Brother)	1 Corinthians 15:7
Paul (post-ascension)	Acts 9:1-9, 1 Corinthians 15:8
Seven Apostles	John 21:1-25
Ten Apostles	Luke 24:36-49, John 20:19-23
Eleven Apostles	John 20:24-31
All Apostles (Great Commission)	Matthew 28:16-20, Mark 16:14-18
All Apostles (pre-ascension)	Acts 1:4-8
Mary Magdalene	John 20:10-18
with Other Women	Matthew 28:1-10
Two Emmaus Disciples	Luke 24:13-35
Over Five Hundred Disciples	1 Corinthians 15:6

[3] Green William H. *General Introduction to the Old Testament – The Text C.* (1899) Scribner's Sons. NY

[4] Gribbin J. & Rees M. *Cosmic Coincidences (1989)* Bantam Books, New York, p.26

Barrow J. & Tipler F. *The Anthropic Cosmological Principle* (1986) Oxford University Press, Oxford, p.367-387

Chapter 2

[1] Gribbin J. & Rees M. *Cosmic Coincidences (1989)* Bantam Books, New York

[2] Barrow J. & Tipler F. *The Anthropic Cosmological Principle* (1986) Oxford University Press, Oxford

[3] Barrow J. & Tipler F. ibid. p.367

Chapter 3

[1] Descartes *Discourse on Method*, (1637)

Chapter 4

[1] One light-year is the distance light travels in one year, about 5.88 thousand billion miles

[2] Materialism asserts: 'All that exists is physical matter and its motion.' Although voiced in various ways, theism's ultimate claim is *something* exists beyond matter and motion. That *something* (God) holds all things (physical matter and its motion) in being.

Chapter 5

[1] Iwamoto N. et. al., The First Chemical Enrichment in the Universe and the Formation of Hyper Metal-Poor Stars, Science 309:451, 15 July 2005

[2] These examples have been vastly oversimplified in order to convey the underlying concepts to the widest possible audience.

[3] Gange, R. *Origins and Destiny* (1986), Ch.3 Did a Supernatural Explosion Create the World?, Word Publishing, Dallas, TX

Chapter 6

[1] Heat energy in a plasma, gas, liquid or solid is associated with the motion of the molecules that comprise the material substance. The faster the motion, the higher is the temperature.

Chapter 7

[1] Each photon has a wavelength and frequency, and travels at the speed of light. Infrared energy has a

frequency just below that of visible light, whereas ultraviolet light has higher frequencies. Very high frequencies are associated with X-rays.

Chapter 8

[1] Stapp H. Mind, Matter & Quantum Mechanics (1981) LBL 12631 Lawrence Berkley Laboratory

[2] Wigner E. Mind-Body Question in Quantum Theory & Measurement, Wheeler J. & Zurek W. Ed. (1983) Princeton U. Press (I.12:180)

[3] Eugene Wigner received the Nobel Prize in Physics in 1963 "for his contributions to the theory of the atomic nucleus and the elementary particles, particularly through the discovery and application of fundamental symmetry principles"

[4] Four of Wigner's more prominent honors include the Enrico Fermi Award, Wigner Fellowship Program at Oak Ridge National Laboratory, renaming of Oak Ridge National Laboratory Auditorium his honor, and the Eugene P. Wigner Reactor Physicist Award (American Nuclear Society)

[5] Waves go up and down (amplitude) with equal distance between adjacent peaks (wavelength) and move through space (velocity). Wave functions are mathematical equations or pictures that are used to describe them.

Chapter 9

[1] Gange R. ibid. Appendix 6

[2] DNA is an acronym for deoxyribonucleic acid — a double helix formed by two strands of nucleotide, each wrapped around the other. Nucleotides are composed of three linked substructures: phosphate, sugar (5-carbon) and a (nitrogenous) side group or 'base.'

[3] Yocky H. A Calculation of the Probability of Spontaneous Biogenesis by Information Theory, (1977) J. Theor. Biol. 67:377

[4] Gange R. ibid. Chap. 10, p.75
[5] Gange R. ibid. Chap. 10, p.76-77
[6] Marcel Golay calculated 220 bits, *Anal. Chem.* (June, 1961) 33:23A); I calculated a low of 235 bits, a high of 293 bits, & a most likely value of 270 bits, *Origins & Destiny*, Appendix 6, (1986), Irvine, TX, Word Books). Jim Trefil also quotes 270 bits, *Space Time Infinity*, (1985) Washington, D.C., Smithsonian Books). Barrow and Tipler calculate a maximum of about 330 bits (*The Anthropic Cosmological Principle* (1986) Oxford University Press, Oxford, p.658-664
[7] Hobson, A. ibid.
[8] Fox, K. The Big Bang Theory: What it is, Where It Came From, and Why it Works (2001) John Wiley & Sons, New York, NY
[9] Thermodynamic closure should not be confused with 'cosmological closure.' The latter speaks to whether or not our universe is gravitationally 'open' (will expand forever), 'fixed' (neither expand nor contact), or gravitationally 'closed' (will cease expanding, and begin to contract)
[10] Hobson, A., Concepts in Statistical Mechanics (1971), Gordon & Breach, NY; Robertson, B., Phys. Rev. (1966), 144:151; (1967) 160:175; Schwegler, Z., Naturforsch (1965) 20a:1543; Jaynes, E., Statistical Physics (1963) (1962 Brandeis Lectures), Ford, K. Ed., New York, NY, W. Benjamin; Zubarev, D., Doklady (1962) 6:776; Scalapino, D., Irreversible Statistical Mechanics and the Principle of Maximum Entropy, (1961), Ph.D. Dissertation, Stanford Univ.;Kawasaki, K., Prog. Theor. Phys., (1960) 23:754;Mori, H., Jour. Phys. Soc. Japan, (1956) 11:1029.

Chapter 10
[1] M. Gell-Mann (1964) *Phys. Let.* 8(3):214
[2] G. Zweig (1964) *CERN Report 8181* / Th 8419
Chapter 11
[1] Heisenberg W. *Physics and Philosophy* (1959) Harper and Row, New York p.160
[2] Wigner E. (1963) *Amer. Jour. Physics* Vol.31 p.6
Chapter 12
[1] Genesis 2:7 God formed man out of the dust of the earth (body)
Breathed into his nostrils the breath of life (spirit)
And man became a living being (soul or person)
Chapter 13
[1] MacDougall, D. (April 1907) *American Medicine*, Vol. 2, No. 4,
Hypothesis Concerning Soul Substance Together with Experimental Evidence of The Existence of Such Substance
[2] MacDougall, D. (May 1907) *Journal of the American Society for Psychical Research,* Vol. I, No. 5
Hypothesis Concerning Soul Substance Together with Experimental Evidence of The Existence of Such Substance
Chapter 14
[1] MacDougall D. ibid. *Journal of the American Society for Psychical Research*
[2] MacDougall D. ibid. *American Medicine*
Chapter 15
[1] Blanke, O. et al. *Brain* Out of body experience and autoscopy of neurological origin, Vol. 127, Issue 2, p.243-258
[2] Holden, J. et al. Eds, *The Field of Near Death Studies Past, Present and Future* (June 2009) Handbook of

Dr. Robert Gange

Near Death Experiences: Thirty Years of Investigation, Greenwood Publishing Group, Westport CT p.1-16

AUTHOR INFORMATION

Robert Gange is a research scientist, NJ State certified professional engineer, and an adjunct professor. He was on staff for over 25 years at the David Sarnoff Research Center in Princeton, New Jersey. His experience embraces guided missiles, laser technology, holographic memories, hybrid computer systems, image processing, cryoelectric devices and integrated electron gun technology.

Educated in five universities (Ph.D. 1978 for extensive research on the application of cryophysics to information storage and retrieval systems), he has made pioneering contributions in several scientific fields, has published numerous papers and is a member of several professional societies including the American Association for the Advancement of Science, and the New York and New Jersey academies of science.

Dr. Gange has received many corporate awards for outstanding achievement both in Science and in Engineering, has been personally honored on nine separate occasions by the National Aeronautics and Space Administration (NASA), and holds over thirty base patents with foreign filings in over twenty-three countries.

19620886R00055

Made in the USA
Charleston, SC
03 June 2013